CHAPTER 1

Adaptation by Natural Selection

Explaining Adaptation before Darwin
Darwin's Theory of Adaptation
    Darwin's Postulates
    An Example of Adaptation by Natural Selection
    Individual Selection
The Evolution of Complex Adaptations
    Why Small Variations Are Important
    Why Intermediate Steps Are Favored by Selection
Rates of Evolutionary Change
Darwin's Difficulties Explaining Variation

Explaining Adaptation before Darwin

Animals and plants are adapted to their conditions in subtle and marvelous ways.

Even the casual observer can see that organisms are well suited to their circumstances. For example, fish are clearly designed for life under the water, and certain flowers are designed to be pollinated by particular species of insects. More careful study reveals that organisms are more than just suited to their environments—they are complex machines, made up of many exquisitely constructed components, or **adaptations,** that interact to help the organism survive and reproduce.

The human eye provides a good example of an adaptation. Eyes are amazingly useful: they allow us to move confidently through the environment, to locate critical resources like food and mates, and to avoid dangers, like predators and cliffs. Eyes are extremely complex structures made up of many interdependent parts (Figure 1.1). Light enters the eye through a transparent opening, then passes through a diaphragm called the iris, which regulates the amount of light entering the eye and allows the

ADAPTATION BY NATURAL SELECTION

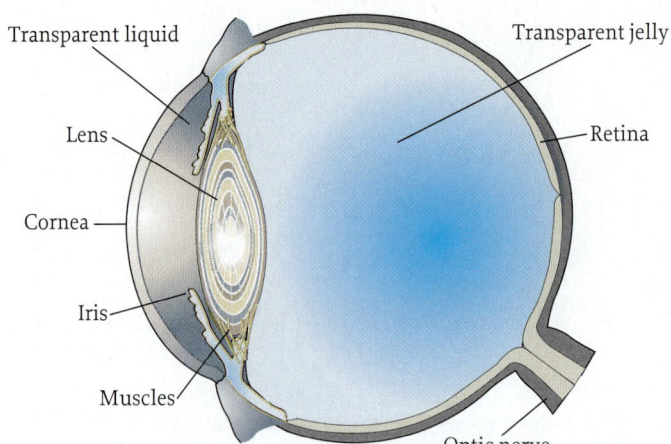

FIGURE 1.1

A cross section of the human eye.

eye to function in a wide range of lighting conditions. The light then passes through a lens that projects a focused image on the retina on the back surface of the eye. Several different kinds of light-sensitive cells then convert the image into nervous impulses that encode information about spatial patterns of color and intensity. These cells are more sensitive to light than the best photographic film. The detailed construction of each of these parts of the eye makes sense in terms of the eye's function—seeing. If we probed into any of these parts, we would see that they too are made of complicated, interacting components whose structure is understandable in terms of their function.

Moreover, differences between human eyes and the eyes of other animals make sense in terms of the types of problems each creature faces. Consider, for example, the eyes of fish and humans (Figure 1.2). The lens in the eyes of humans and other terrestrial mammals is much like a camera lens; it is shaped like a squashed football and has the same index of refraction (a measure of light-bending capacity) throughout. In contrast, the lens in fish eyes is a sphere located at the center of the curvature of the retina, and the index of refraction of the lens increases smoothly from the surface of the lens to the center. It turns out that this kind of lens, called a spherical gradient lens, provides a sharp image over a full 180° visual field, a very short focal length, and high light-gathering power—all desirable properties. Terrestrial creatures like us cannot use this design because light is bent when it passes from the air through the cornea, the transparent cover of the pupil, and this fact constrains the design of the remaining

FIGURE 1.2

Like those of other terrestrial mammals, human eyes have more than one light-bending element. A ray of light entering the eye (dashed lines) is bent first as it moves from the air to the cornea and then again as it enters and leaves the lens. In contrast, fish eyes have a single lens that bends the light throughout its volume. As a result, fish eyes have a short focal length and high light-gathering power.

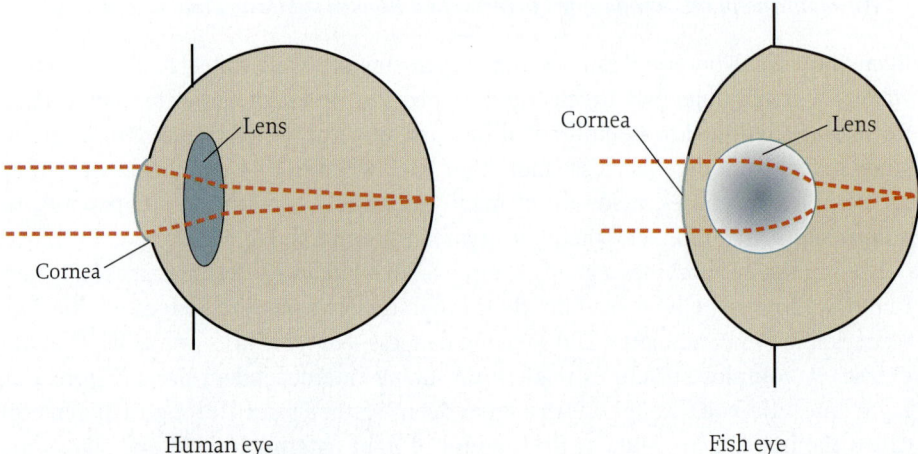

Darwin's Theory of Adaptation

lens elements. In contrast, light is not bent when it passes from water through the cornea of aquatic animals, and the design of their eyes takes advantage of this fact.

Before Darwin there was no scientific explanation for the fact that organisms are well adapted to their circumstances.

As many 19th-century thinkers were keenly aware, complex adaptations like the eye demand a different kind of explanation than other natural objects do. This is not simply because adaptations are complex, since many other complicated objects exist in nature. Adaptations require a special kind of explanation because they are complex in a particular, highly improbable way. For example, the Grand Canyon, with its maze of delicate towers intricately painted in shades of pink and gold, is byzantine in its complexity (Figure 1.3). However, given a different geological history, the Grand Canyon might be quite different—different towers, in different hues—yet we would still recognize it as a canyon. The particular arrangement of painted towers of the Grand Canyon is improbable, but the existence of a spectacular canyon with a complex array of colorful cliffs in the dry sandstone country of the American Southwest is not unexpected at all, and, in fact, wind and water produced several other canyons in this region. In contrast, any substantial changes in the structure of the eye would prevent the eye from functioning, and then we would no longer recognize it as an eye. If the cornea was opaque, or the lens was on the wrong side of the retina, then the eye would not transmit visual images to the brain. It is highly improbable that natural processes would randomly bring together bits of matter having the detailed structure of the eye because only an infinitesimal fraction of all arrangements of matter would be recognizable as a functioning eye.

In Darwin's day, most people were not troubled by this problem because they believed that adaptations were the result of divine creation. In fact, the theologian William Paley used a discussion of the human eye to argue for the existence of God in his book *Natural Theology*, published in 1802. Paley argued that the eye is clearly *designed* for seeing, and where there is design in the natural world there certainly must be a heavenly designer.

While most scientists were satisfied with this reasoning, a few, including Charles Darwin, sought other explanations.

FIGURE 1.3

The Grand Canyon is an impressive geological feature, but much less remarkable in its complexity than the eye.

Darwin's Theory of Adaptation

Charles Darwin was expected to become a doctor or clergyman, but instead revolutionized science.

Charles Darwin was born into a well-to-do, intellectual, and politically liberal family in England. Like many prosperous men of his time, Darwin's father wanted his son to become a doctor. But after failing at the prestigious medical school at the University of Edinburgh, Charles went on to Cambridge University, resigned to becoming a country parson. He was, for the most part, an undistinguished student, and was much

FIGURE 1.4
The HMS *Beagle* in Beagle Channel on the southern coast of Tierra del Fuego.

more interested in tramping through the fields around Cambridge in search of beetles than in studying Greek and mathematics. After graduation, one of Darwin's biology professors, William Henslow, provided him with a chance to pursue his passion for natural history as a naturalist on the HMS *Beagle*.

The *Beagle* was a Royal Navy vessel whose charter was to spend two to three years mapping the coast of South America and then to return to London, perhaps by circling the globe (Figure 1.4). Darwin's father forbade him to go, preferring that he get serious about his career in the clergy, but Darwin's uncle (and future father-in-law) Josiah Wedgwood intervened. The voyage turned out to be the turning point in Darwin's life. His work during the voyage established his reputation as a skilled naturalist. His observations of living and fossil animals ultimately convinced him that plants and animals sometimes change slowly through time, and that such evolutionary change was the key to understanding how new species came into existence. This view was rejected by most scientists of the time and was considered heretical by the general public.

Darwin's Postulates

Darwin's theory of adaptation follows from three postulates: the struggle for existence, variation in fitness, and the inheritance of variation.

In 1838, shortly after the *Beagle* returned to London, Darwin formulated a simple mechanistic explanation for *how* species change through time. His theory follows from three postulates:

1. The ability of a population to expand is infinite, but the ability of any environment to support populations is always finite.
2. Organisms within populations vary, and this variation affects the ability of individuals to survive and reproduce.
3. The variations are transmitted from parents to offspring.

Darwin's first postulate means that populations grow until they are checked by the dwindling supply of resources in the environment. Darwin referred to the resulting competition for resources as "the struggle for existence." For example, animals require food to grow and reproduce. When food is plentiful, animal populations grow until their numbers exceed the local food supply. Since resources are always finite, it follows that not all individuals in a population will be able to survive and reproduce. According to the second postulate, some individuals will possess traits that enable them to survive and reproduce more successfully (producing more offspring) than others in the same environment. The third postulate holds that if the advantageous traits are inherited by offspring, then these traits will become more common in succeeding generations. Thus, traits that confer advantages in survival and reproduction are retained in the population, and traits that are disadvantageous disappear.

DARWIN'S THEORY OF ADAPTATION

(a)

(b)

FIGURE 1.5
(a) The islands of the Galápagos, which are located off the coast of Ecuador, house a variety of unique species of plants and animals. (b) Cactus finches from Charles Darwin, *The Zoology of the Voyage of H.M.S. Beagle.*

When Darwin coined the term **natural selection** for this process, he was making a deliberate analogy to the artificial selection practiced by animal and plant breeders of his day. A much more apt term would be "evolution by variation and selective retention."

AN EXAMPLE OF ADAPTATION BY NATURAL SELECTION

Contemporary observations of Darwin's finches provide a particularly good example of how natural selection produces adaptations.

In his autobiography, first published in 1887, Darwin claimed that the curious pattern of adaptations he observed among the several species of finches that live on the Galápagos Islands off the coast of Ecuador—now referred to as Darwin's finches—was crucial in the development of his ideas about evolution (Figure 1.5). Recently discovered documents suggest that Darwin was actually quite confused about the Galápagos finches during his visit, and they played little role in his discovery of natural selection. Nonetheless, Darwin's finches hold a special place in the minds of most biologists.

Peter and Rosemary Grant, biologists at Princeton University, conducted a landmark study of the ecology and evolution of one particular species of Darwin's finches on one of the Galápagos Islands. The study is remarkable because the Grants were able to directly document how Darwin's three postulates lead to evolutionary change. The island, Daphne Major, is home to the medium ground finch *(Geospiza fortis)*, a small bird that subsists mainly by eating seeds (Figure 1.6). The Grants and their colleagues caught, measured, weighed, and banded nearly every finch on the island each year, some 1500 birds in all. They also kept track of critical features of the birds' environment, such as the distribution of seeds of various sizes, and observed the birds' behavior. A few years into the Grants' study, a severe drought struck Daphne

FIGURE 1.6
The medium ground finch, *Geospiza fortis,* uses its beak to crack open seeds. (Photograph courtesy of Peter Grant.)

(a) (b)

FIGURE 1.7
(a) This is how Daphne Major looked in March 1976, after a year of good rains.
(b) This is how Daphne Major looked in March 1977, after a year of very little rain.

Major (Figure 1.7). During the drought, plants produced many fewer seeds, and the finches soon depleted the stock of small, soft, easily processed seeds, leaving only large, hard, difficult-to-process seeds (Figure 1.8). The bands on the birds' legs enabled the Grants to track the fates of individual birds during the drought, and the regular measurements that they had made of the birds allowed them to compare the traits of birds that survived the drought with the traits of those that perished. They also made detailed records of the environmental conditions, which allowed them to determine how the drought affected the finch's habitat. It was this vast body of data that enabled the Grants to document the action of natural selection among the finches of Daphne Major.

The Grants' data show how the processes identified in Darwin's postulates lead to adaptation.

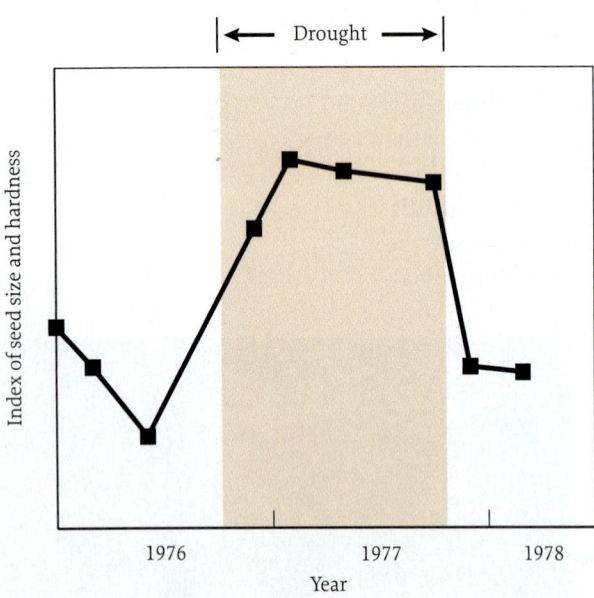

FIGURE 1.8
During the two-year drought, the size and hardness of seeds available on Daphne Major increased because birds consumed all of the desirable small, soft seeds, leaving mainly larger and harder seeds. Each point represents an index of seed size and hardness at a given time.

The events on Daphne Major embodied all three of Darwin's postulates. First, the supply of food on the island was not sufficient to feed the entire population, and many finches did not survive the drought. From the beginning of the drought in 1976 until the rains came nearly two years later, the population of medium ground finches on Daphne Major declined from 1200 birds to only 180.

Second, beak depth (the top-to-bottom dimension of the beak) varied among the birds on the island, and this variation affected the birds' survival. Before the drought began, the Grants and their colleagues had observed that birds with deeper beaks were able to process large, hard seeds more easily than birds with shallower beaks were. Deep-beaked birds usually concentrated on large seeds while shallow-beaked birds normally focused their efforts on small seeds. The open bars in the upper histogram in Figure 1.9b show what the distribution of beak sizes in the population was like before the drought. The height of each open bar represents the number of birds with beaks in a given range of depths, for example, 8.8 to 9.0 mm or 9.0 to 9.2 mm. During the drought, the relative abundance of small seeds decreased, forcing shallow-beaked birds to shift to larger and harder seeds. Shallow-beaked birds

Studying Evolution in Darwin's Finches

Half past seven on Daphne Major. Peter and Rosemary Grant sit themselves down on stones, a few steps from their traps. Peter opens a yellow notebook with waterproof pages. "Okay," he says. "Today is the twenty-fifth."

It is the twenty-fifth of January, 1991. There are four hundred finches on the island at this moment, and the Grants know every one of the birds on sight, the way shepherds can tell every sheep in their flocks. In other years there have been more than a thousand finches on Daphne Major, and Peter and Rosemary could still recognize each one. The flock was down to three hundred once. The number is falling toward that now. The birds have gotten less than a fifth of an inch of rain in the last forty-four months: in 1,320 days, 5 millimeters of rain.

The Grants, and the Grants' young daughters, and a long line of assistants, keep coming back to this desert island like sentries on a watch. They have been observing Daphne Major for almost two decades, or about twenty generations of finches. By now Peter and Rosemary Grant know many of the birds' family trees by heart—again like shepherds, or like Bible scholars, who know that Abraham begat Isaac, and Isaac begat Jacob; and Abraham also begat Jokshan, who begat Dedan, who begat Asshurim, Letushim, and Leummim.

In each generation there are always a few birds, just one or two in a hundred, that keep away from the Grants and refuse to be caught. This morning Rosemary, after a week of watching and plotting, has just captured two of the wariest, most difficult finches on the island. She caught them both in the space of a single minute, high on the island's north rim, next to a fallen cactus pad, in black box traps baited with green bananas. "How about that," she cried, when the traps' doors clicked shut. And when Peter strode through the cactus trees and across the lava rubble to join her, Rosemary lifted up her first prize, fluttering in a blue pouch. "I deserve a bottle of wine for this!"

Now the Grants are sitting beside the traps at the edge of a cliff, 100 meters above the Pacific Ocean. Except for the honking and whistling of two masked boobies, courting on a rock nearby, the scene is quiet. The ocean is more than pacific; it is flat as a pond. The morning's weather is what Charles Darwin described in his diary when he first saw the Galápagos archipelago, "a steady, gentle breeze of wind & gloomy sky."

From the upper rim of Daphne Major, on clearer mornings than this one, Rosemary and Peter can see the island of Santiago, where Darwin camped for nine days. They can also see the island of Isabela, where Darwin spent one day. They can make out more than a dozen other islands and black lava ruins that Darwin never had a chance to visit, including an islet known officially as Sin Nombre (that is, Nameless) and another black speck called Eden.

"If I have seen further," Isaac Newton once wrote, with celebrated modesty, "it is by standing upon the shoulders of Giants." The dark volcanoes of the Galápagos are Darwin's shoulders. These islands meant more to him than any other stop in his five-year voyage around the world. "Origin of all my views," he called them once—the origin of the *Origin of Species*. The Grants are doing what Darwin could not do, going back to the Galápagos year after year; and the Grants are seeing there what Darwin did not imagine could be seen at all.

Rosemary unlatches their tool kit, a tackle box. From it, Peter extracts a pair of jeweler's spectacles, a plastic mask with bulging lenses, which make him look like Robinson Crusoe from Mars. "Okay, Famous Bird," Peter says. "*Ow!* Famous Bird has decided to bite the hand that feeds him." He grasps the finch with one hand, and its head sticks out observantly from his fist. The bird is about the size of a sparrow, and jet-black, with a black beak and shiny dark eyes.

Rosemary hands Peter a pair of calipers. "Now, here we go," Peter says. "Wing length, 72 millimeters."

Rosemary jots the number in the yellow notebook.

"Tarsus length, 21.5." (The tarsus is the bird's leg.)

Rosemary writes it down.

"Beak length, 14.9 millimeters," Peter recites. "Beak depth, 8.8. Beak width, 8 millimeters."

"Black Five plumage." The Grants rate the birds' plumage from zero, which is brown, to five, totally black. Black Five means a mature male.

"Beak black." Normally these birds' beaks are pale, the color of horn. A black beak means the bird is ready to mate.

Peter dangles the bird in a little weighing cup. "Weight, 22.2 grams."

"This bird has lived a long time," he muses. "Thirteen years." There are only three others of its generation still alive on the island, and none older. "But I don't think there's a single one of his offspring flying around. Not *one* has made it to the breeding season." The bird has been a father many times, and never once a grandfather.

Peter puts a gray ring and a brown ring on the bird's left ankle. He puts a light green ring over a metal one on its right ankle. Bands like these, and an ingenious color code, help the Grant team to keep track of their flocks from dawn to dusk, from the cliffs at the base of the island to this guano-painted rubble at the rim.

Peter holds the bird in his fist one more time and inspects its beak in profile. In rushing up to join Rosemary at the rim, he has forgotten his camera. Otherwise he would photograph the bird just so, from a distance of 27 centimeters. That is the Grants' standard mug shot for one of Darwin's finches.

SOURCE: From pp. 3–6 in *The Beak of the Finch,* by Jonathan Weiner. Copyright © 1994 by Jonathan Weiner. Reprinted by permission of Alfred A. Knopf, Inc.

were then at a distinct disadvantage because it was harder for them to crack these seeds. The distribution of individuals within the population changed during the drought because finches with deeper beaks were more likely to survive than finches with shallow beaks (Figure 1.9a). The shaded bars of the lower histogram in Figure 1.9b show what the distribution of beak depths would have been like among the sur-

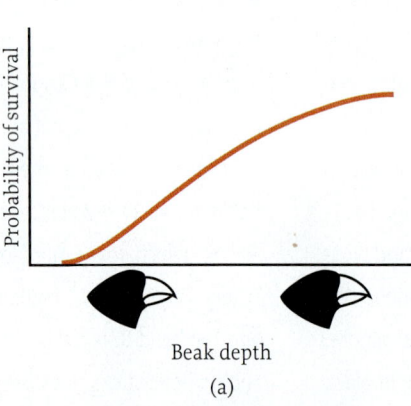
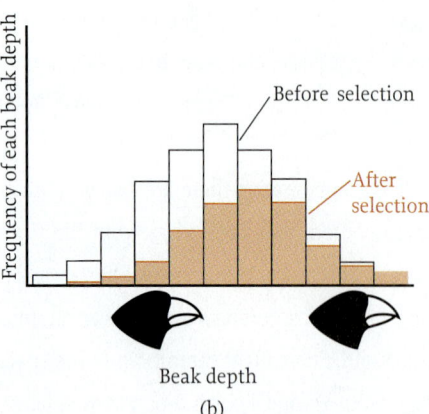

Beak depth
(a)

Beak depth
(b)

FIGURE 1.9

A schematic diagram of how directional selection increased mean beak depth among medium ground finches on Daphne Major. (a) The probability of survival for birds of different beak depths is plotted. Birds with shallow beaks are less likely to survive than are birds with deep beaks. (b) The height of each bar represents the numbers of birds whose beak depths fall within each of the intervals plotted on the x-axis, with beak depth increasing to the right. The histogram with open bars shows the distribution of beak depths before the drought began. The histogram with shaded bars shows the distribution of beak depths after a year of drought. Notice that the number of birds in each category has decreased. Since birds with deep beaks were less likely to die than birds with shallow beaks, the peak of the distribution has shifted to the right, indicating that the mean beak depth has increased.

vivors. Because many birds died, there were fewer remaining in each category. However, mortality was quite specific. The proportion of shallow-beaked birds that died greatly exceeded the proportion of deep-beaked birds that died. As a result, the lower histogram shows a shift to the right, which means that the average beak depth in the population increased. Thus, the average beak depth among the survivors of the drought is greater than the average beak depth in the same population before the drought.

Third, parents and offspring had similar beak depths. The Grants discovered this by capturing and banding nestlings and recording the identity of the nestlings' parents. When the nestlings became adults, the Grants recaptured and measured them. When compiling these data, the Grants discovered that parents with deep beaks on average produced offspring with deep beaks (Figure 1.10). Since parents were drawn from the pool of individuals who survived the drought, their beaks were on average deeper than those of the original residents of the island, and since offspring resemble their parents, the average beak depth of the survivors' offspring was greater than the average beak depth before the drought. This means that through natural selection, the average **morphology** (an organism's size, shape, and composition) of the bird population changed so that birds became better adapted to their environment. This process, operating over approximately two years, led to a 4% increase in the mean beak depth in this population (Figure 1.11).

 Selection preserves the status quo when the most common type is the best adapted.

So far, we have seen how natural selection led to adaptation as the population of finches on Daphne Major evolved in response to changes in their environment. Will this process continue forever? If it did, eventually all the finches would have deep enough beaks to efficiently process the largest seeds available. However, large beaks have disadvantages as well as benefits. The Grants showed, for instance, that birds with large beaks are less likely to survive the juvenile period than are birds with small beaks, probably because they require more food (Figure 1.12). Evolutionary theory predicts that over time, selection will increase the average beak depth in the population until the costs of larger-than-average beak size exceed the benefits. At this point, finches with the average beak size in the population will be the most likely to survive

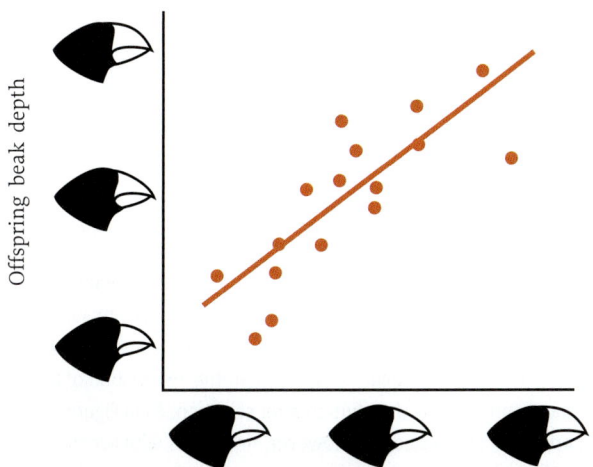

FIGURE 1.10

Parents with deeper-than-average beaks tend to have offspring with deeper-than-average beaks. Each point represents one offspring. Offspring beak depth is plotted on the vertical axis (deeper beaks farther up the axis), and the average of the two parents' beak depth is plotted on the horizontal axis (deeper beaks farther to the right).

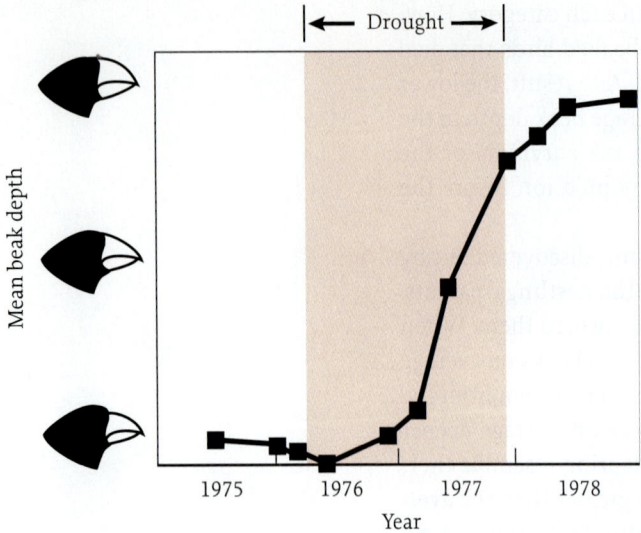

FIGURE 1.11

The average beak depth in the population of medium ground finches on Daphne Major increased during the drought of 1975–1978. Each point plots an index of average beak depth of the population in a particular year. Deeper beaks are plotted higher on the y-axis.

and reproduce, and finches with deeper or shallower beaks than the new average will be at a disadvantage. When this is true, beak size does not change, and we say that an **equilibrium** exists in the population with regard to beak size. The process that produces this equilibrium state is called **stabilizing selection.** Notice that even though the average characteristics of the beak in the population will not change in this situation, selection is still going on. Selection is required to change a population, and selection is also required to keep a population the same.

It might seem that beak depth would also remain unchanged if this trait had no effect on survival (or put another way, if there were no selection favoring one type of beak over another). Then, all types of birds would be equally likely to survive from one generation to the next, and beak depth would remain constant. This logic would be valid if selection were the only process affecting beak size. However, real populations are also affected by other processes that cause **traits,** or **characters,** to change in unpredictable ways. We will discuss these processes further in Chapter 3. The point to remember here is that populations do not remain static over the long run unless selection is operating.

 Species are populations of varied individuals that may or may not change through time.

As the Grants' work on Daphne Major makes clear, a species is not a fixed type or entity. Species change in their general characteristics from generation to generation

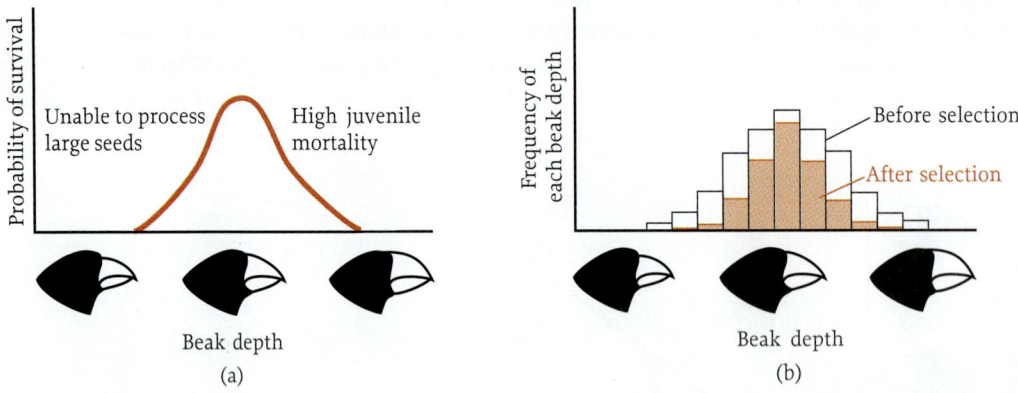

FIGURE 1.12

When birds with the most common beak depth are most likely to survive and reproduce, natural selection keeps the mean beak depth constant. (a) Birds with deep or shallow beaks are less likely to survive than birds with average beaks. Birds with shallow beaks cannot process large, hard seeds, and birds with deep beaks are less likely to survive to adulthood. In (b) the open bars represent the distribution of beak depths before selection, and the shaded bars represent the distribution after selection. As in Figure 1.9, notice that there are fewer birds in the population after selection. However, since birds with average beaks are most likely to survive, the peak of the distribution of beak depths is not shifted and mean beak depth remains unchanged.

according to the postulates Darwin described. Before Darwin, however, people thought of species as fixed, unchanging categories much the same way that we think of geometrical figures. A finch could no more change its properties than a triangle could. If a triangle acquires another side, it is not a modified triangle but rather a rectangle. In much the same way, to biologists before Darwin, a changed finch is not a finch at all. Ernst Mayr, the distinguished evolutionary biologist, calls this pre-Darwinian view of immutable species, essentialism. According to Darwin's theory, a **species** is a dynamic *population* of individuals. If the characteristics of a particular species appear static over a long period of time, it is because this type of individual is consistently favored by stabilizing selection. Both stasis (staying the same) and change result from natural selection, and both require explanation in terms of natural selection. Stasis is not the natural state of species.

Individual Selection

 Adaptation results from the competition among individuals, not between entire populations or species.

It is important to notice that selection produces adaptations that benefit *individuals*. Such adaptation may or may not benefit the population or species. In the case of simple morphological characters such as beak depth, selection probably does allow the population of finches to compete more effectively with other populations of seed predators. However, this need not be the case. Selection often leads to changes in behavior or morphology that increase the reproductive success of individuals but decrease the average reproductive success of the group, population, and species.

The fact that almost all organisms produce many more offspring than are necessary to maintain the species provides an example of the conflict between individual and group interests. A female monkey may, on average, produce 10 offspring during her lifetime (Figure 1.13). In a stable population, perhaps only two of these offspring will survive and reproduce. From the point of view of the species, the other eight are a waste of resources. They compete with other members of their species for food, water, and sleeping sites. The demands of a growing population can lead to serious overexploitation of the environment, and the species as a whole might be more likely to survive if all females produced fewer offspring. However, this does not happen because natural selection among individuals favors females who produce many offspring.

FIGURE 1.13
A female blue monkey holds her infant. (Photograph courtesy of Marina Cords.)

To see why selection on individuals will lead to this result, let's consider a simple, hypothetical case. Suppose the females of a particular species of monkey are maximizing individual reproductive success when they produce 10 offspring. Females who produce more than or less than 10 offspring will tend to leave fewer descendants in the next generation. Further suppose that the likelihood of the species becoming extinct would be lowest if females produced only two offspring apiece. Now suppose that there are two kinds of females. Most of the population is composed of low-fecundity females who produce just two offspring each, but there are a few high-fecundity females who produce 10 offspring each. (**Fecundity** is the term demographers use for the ability to produce offspring.) High-fecundity females have high-fecundity daughters, and low-fecundity females have low-fecundity daughters. The proportion of high-fecundity females will increase in the next generation because such females produce more offspring than do low-fecundity females. Over time, the proportion of high-fecundity females in the population will increase rapidly. As fecundity increases, the population will grow rapidly and may exhaust available resources. This in turn will increase the chance that the species becomes extinct. However, this fact is irrelevant to the evolution of fecundity before the extinction, because natural selection results from competition among individuals, not competition among species.

The idea that natural selection operates at the level of the individual is a key element in understanding adaptation. In discussing the evolution of social behavior in Chapter 8, we will encounter several additional examples of situations in which selection increases individual success but decreases the competitive ability of the population.

The Evolution of Complex Adaptations

The example of the evolution of beak depth in the medium ground finch illustrates how natural selection can cause adaptive change to occur rapidly in a population. Deeper beaks enabled the birds to survive better, and deeper beaks soon came to predominate in the population. Beak depth is a fairly simple character, lacking the intricate complexity of an eye. However, as we will see, the accumulation of small variations by natural selection can also give rise to complex adaptations.

WHY SMALL VARIATIONS ARE IMPORTANT

There are two categories of variation: continuous and discontinuous.

It was known in Darwin's day that most variation is continuous. An example of **continuous variation** is the distribution of heights in people. Humans grade smoothly from one extreme to the other (short to tall), with all the intermediate types (in this case, heights) represented. However, Darwin's contemporaries also knew about **discontinuous variation,** in which a number of distinct types exist with no intermediates. In humans, height is also subject to discontinuous variation. For example, there is a genetic condition called **achondroplasia,** which causes affected individuals to be much shorter than other people, have proportionately shorter arms and legs, and bear a variety of other distinctive features. Discontinuous variants are usually quite rare in nature. Nonetheless, many of Darwin's contemporaries, who were convinced of the reality of evolution, believed that new species arise as discontinuous variants.

Discontinuous variation is not important for the evolution of complex adaptations because complex adaptations are extremely unlikely to arise in a single jump.

Unlike most of his contemporaries, Darwin thought that discontinuous variation did not play an important role in evolution. A hypothetical example, described by Oxford University biologist Richard Dawkins in his book *The Blind Watchmaker,* illustrates Darwin's reasoning. Dawkins recalls an old story in which an imaginary collection of monkeys sit at typewriters happily typing away. Lacking the ability to read or write, they strike keys at random. Given enough time, the story goes, the monkeys will reproduce all the great works of Shakespeare. Dawkins points out that this is not likely to happen in the lifetime of the universe, let alone the lifetime of one of the monkey typists. To illustrate why it would take so long, Dawkins presents these illiterate monkeys with a much simpler problem: reproducing a single line from *Hamlet,* "Methinks it is like a weasel." To make the problem even simpler for the monkeys, Dawkins ignores the difference between uppercase and lowercase letters and omits all punctuation except spaces. There are 28 characters (including spaces) in the phrase. Since there are 26 characters in the alphabet and Dawkins is keeping track of spaces, each time a monkey types a character, there is only a 1-in-27 chance that she will type the right character. There is also only a 1-in-27 chance that the second character will be correct. Again, there is a 1-in-27 chance that the third character will be right, and so on up to the 28th character. Thus the chance that a monkey will type the correct sequence at random is 1/27 times itself 28 times, or

$$\underbrace{\frac{1}{27} \times \frac{1}{27} \times \frac{1}{27} \times \cdots \times \frac{1}{27}}_{28 \text{ times}} \approx 10^{-40}$$

This is a *very* small number. To get a feeling for how small a chance there is of the monkeys typing the sentence correctly, suppose a very fast computer (faster than any currently in existence) could generate 100 billion (10^{11}) characters per second and run for the lifetime of the earth, about 4 billion years, or 10^{17} seconds. Then, the chance of randomly typing the line "Methinks it is like a weasel" even once during the whole of the earth's history would be about 1 in a trillion! Typing the whole play is obviously astronomically less likely, and although *Hamlet* is a very complicated thing, it is much less complicated than a human eye. There's no chance that a structure like the human eye would arise by chance in a single trial. If it did, it would be as the astrophysicist Sir Fred Hoyle said, "like a hurricane blowing through a junk yard and chancing to assemble a Boeing 747."

Complex adaptations can arise through the accumulation of small random variations by natural selection.

Darwin argued that continuous variation is essential for the evolution of complex adaptations. Once again, Richard Dawkins provides an example that makes clear Darwin's reasoning. Again imagine a room full of monkeys and typewriters, but now the rules of the game are different. The monkeys type the first 28 characters at random, and then during the next round they attempt to copy the same initial string of letters and spaces. Most of the sentences are just copies of the previous string, but because monkeys sometimes make mistakes, some strings have small variations, usually in only a single letter. During each trial, the monkey trainer selects the string that most resembles Shakespeare's phrase "Methinks it is like a weasel" as the string to be copied by all the monkeys in the next trial. This process is repeated until the monkeys come up with the correct string. Calculating the exact number of trials required to generate the correct sequence of characters is quite difficult, but it is easy to simulate the process on a computer. Here's what happened when Dawkins performed the simulation. The initial random string is

WDLMNLT DTJBKWIRZREZLMQCO P

After one trial Dawkins got

WDLMNLT DTJBSWIRZREZLMQCO P

After 10 trials:

MDLDMNLS ITJISWHRZREZ MECS P

After 20 trials:

MELDINLS IT ISWPRKE Z WECSEL

After 30 trials:

METHINGS IT ISWLIKE B WECSEL

After 40 trials:

METHINKS IT IS LIKE I WEASEL

The exact phrase was reached after 43 trials. Dawkins reports that it took his 1985-vintage Macintosh only 11 seconds to complete this task.

Selection can give rise to great complexity starting with small random variations because it is a *cumulative* process. As the typing monkeys show us, it is spectacularly unlikely that a single random combination of keystrokes will produce the correct sentence. However, there is a much greater chance that some of the many *small* random changes will be advantageous. The combination of reproduction and selection allows the typing monkeys to accumulate these small changes until the desired sentence is reached.

WHY INTERMEDIATE STEPS ARE FAVORED BY SELECTION

The evolution of complex adaptations requires that all of the intermediate steps be favored by selection.

There is a potent objection to the example of the typing monkeys. Natural selection, acting over time, can lead to complex adaptations, but it can do so only if each small change along the way is itself adaptive. While it is easy to assume that this is true in a hypothetical example of character strings, many people have argued that it is unlikely for every one of the changes necessary to assemble a complex organ like the eye to be adaptive. An eye is only useful, it is claimed, once all parts of the complexity are assembled; until then, it is worse than no eye at all. After all, what good is 5% of an eye?

Darwin's answer, based on the many adaptations for seeing or sensing light that exist in the natural world, was that 5% of an eye *is* often better than no eye at all. It is quite possible to imagine that a very large number of small changes—each favored by selection—led cumulatively to the wonderful complexity of the eye. Living mollusks, which display a broad range of light-sensitive organs, provide examples of many of the likely stages in this process (Figure 1.14):

1. Many invertebrates have a simple light-sensitive spot (Figure 1.14a). Photoreceptors of this kind have evolved many times from ordinary epidermal (surface) cells—usually ciliated cells whose biochemical machinery is light-sensitive. Those individuals whose cells are more sensitive to light are favored when information about changes in light intensity is useful. For example, a drop in light intensity may often be an indicator of a predator in the vicinity.
2. By having the light-sensitive cells in a depression, the organism will get some additional information about the direction of the change in light intensity (Figure 1.14a). The surface of organisms is variable, and those individuals whose photoreceptors are in depressions will be favored by selection in environments where

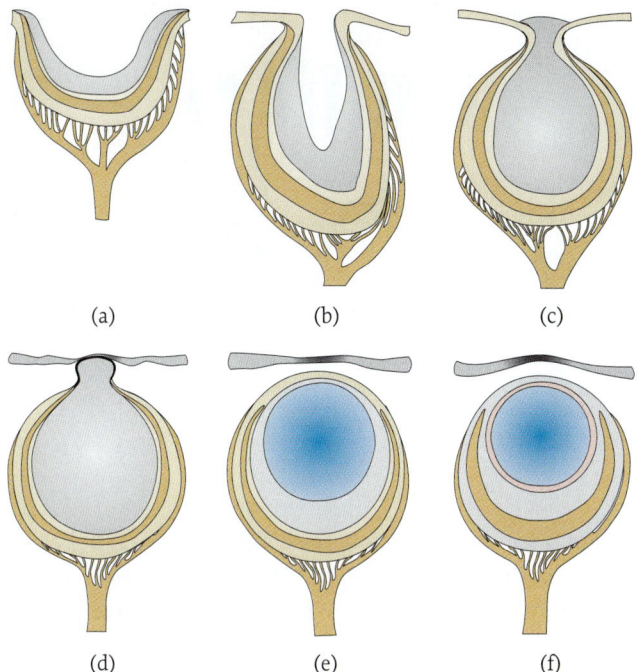

FIGURE 1.14

Living gastropod mollusks illustrate all of the intermediate steps between a simple eye cup and a camera-type eye: (a) The eye pit of a limpet, *Patella* sp.; (b) the eye cup of Beyrich's split shell, *Pleurotomaria beyrichi*; (c) the pinhole eye of a California abalone, *Haliotis* sp.; (d) the closed eye of a turban shell, *Turbo creniferus*; (e) the lens eyes of the spiny dye murex, *Murex brandaris*; (f) the lens eyes of the Atlantic dog whelk, *Nucella lapis*. (Lens is shaded in e and f.)

such information is useful. For example, mobile organisms may need better information about what is happening in front of them than do immobile ones.

3. Through a series of small steps, the depression could get deeper, and each step could be favored by selection because better directional information would be available (Figure 1.14b).
4. If the depression got deep enough, it could form images on the light-sensitive tissue, much the way pinhole cameras form images on photographic film (Figure 1.14c). In settings in which detailed images are useful, selection could then favor the elaboration of the neural machinery necessary to interpret the image.
5. The next step is the formation of a transparent cover (Figure 1.14d). This might be favored because it protects the interior of the eye from parasites and mechanical damage.
6. A lens could evolve either through gradual modification of the transparent cover or through the modification of internal structures within the eye (Figure 1.14e, f).

Notice that evolution produces adaptations like a tinkerer, not an engineer. New organisms are created by small modifications of existing ones, not by starting with a clean slate. Clearly many beneficial adaptations will not arise because they are blocked at some step along the way when a particular variation is not favored by selection. Darwin's theory explains how complex adaptations can arise through natural processes, but it does not predict that every adaptation, or even most adaptations, that could occur have occurred or will occur. This is not the best of all possible worlds; it is just one of many possible worlds.

 Sometimes unrelated species have independently evolved the same complex adaptation. This suggests that the evolution of complex adaptations by natural selection is not a matter of mere chance.

The fact that natural selection constructs complex adaptations like a tinkerer might lead you to think that the assembly of complex adaptations is a chancy busi-

FIGURE 1.15

The marsupial wolf that lived in Tasmania until early in the 20th century (drawn from a photograph of one of the last living animals). Similarities with placental wolves of North America and Eurasia illustrate the power of natural selection to create complex adaptations. Their last common ancestor was probably a small insectivorous creature like the shrew.

(a)

(b)

FIGURE 1.16

Complex eyes with lenses have evolved independently in a number of different kinds of aquatic animals, including this slingjaw wrasse (a) and squid (b).

ness. If even a single step was not favored by selection, the adaptation could not arise. Such reasoning suggests that complex adaptations are mere coincidence. While chance does play a very important role in evolution, the power of cumulative natural selection should not be underestimated. The best evidence that selection is a powerful process for generating complex adaptations comes from a phenomenon called **convergence,** the evolution of similar adaptations in unrelated groups of animals.

The similarity between the marsupial faunas of Australia and South America and the placental faunas of the rest of the world provides a good example of convergence. In most of the world, the mammalian fauna is dominated by **placental mammals,** which nourish their young in the uterus during long pregnancies. However, both Australia and South America became separated from an ancestral supercontinent, known as Pangaea, long before placental mammals evolved. In Australia and South America, **marsupials** (nonplacental mammals, like kangaroos, that rear their young in external pouches) came to dominate the mammalian fauna, filling all available mammalian niches. Some of these marsupial mammals were quite similar to the placental mammals on the other continents. For example, there was a marsupial wolf in Australia that looked very much like placental wolves of Eurasia, even sharing subtle features of their feet and teeth (Figure 1.15). These marsupial wolves became extinct in the 1930s. Similarly, in South America, a marsupial saber-toothed cat independently evolved many of the same adaptations as the placental saber-toothed cat that stalked North America 10,000 years ago. These similarities are more impressive when you consider that the last common ancestor of marsupial and placental mammals was a small, nocturnal, insectivorous creature, something like a shrew, that lived about 120 mya (million years ago). Thus, selection transformed a shrew step by small step, each step favored by selection, into a saber-toothed cat—and it did it twice. This cannot be coincidence.

The evolution of eyes provides another good example of convergence. Remember that the spherical gradient lens is a good lens design for aquatic organisms because it has good light-gathering ability and provides a sharp image over the full 180° visual field. Complex eyes with lenses have evolved independently eight different times in distantly related aquatic organisms: once in fish, once in cephalopod mollusks like squid, several times among gastropod mollusks like Atlantic dog whelk, once in annelid worms, and once in crustaceans (Figure 1.16). These are very diverse creatures whose last common ancestor was a simple creature that did not have a complex eye. Nonetheless, in every case they have evolved very similar spherical gradient lenses. Moreover, no other lens design is found in aquatic animals. Despite the seeming chanciness of assembling complex adaptations, natural selection has achieved the optimal design in every case.

Rates of Evolutionary Change

Natural selection can cause evolutionary change that is much more rapid than we commonly observe in the fossil record.

In Darwin's day, the idea that natural selection could change a chimpanzee into a human, much less that it might do so in just a few million years, was unthinkable. Though people are generally more accepting of evolution today, many still think of evolution by natural selection as a glacially slow process that requires millions of years to accomplish noticeable change. Such people often doubt that there has been enough time for selection to accomplish the evolutionary changes observed in the fossil record. And yet, as we will see in subsequent chapters, most scientists now believe that humans evolved from an apelike creature in only 5 million to 10 million years. In fact, some of the rates of selective change observed in contemporary populations are far faster than necessary for such a transition. The puzzle is not whether there has been enough time for natural selection to produce the adaptations that we observe. The real puzzle is why the change observed in the fossil record was so slow.

The Grants' observation of the evolution of beak morphology in Darwin's finches provides one example of rapid evolutionary change. The medium ground finch of Daphne Major is one of 14 species of finches that live in the Galápagos. Evidence suggests that all 14 are descended from a single South American species that migrated to the newly emerged islands about half a million years ago (Figure 1.17). This doesn't seem like a very long time. Is it possible that natural selection created 14 species in only half a million years?

To start to answer the question, let's calculate how long it would take for the medium ground finch to come to resemble its closest relative, the large ground finch *(Geospiza magnirostris)*, in beak size and weight (Figure 1.18). The large ground finch is 75% heavier than the medium ground finch, and its beak is about 20% deeper. Remember that beak size increased about 4% in two years during the 1977 drought. The Grants' data indicate that body size also increased by a similar amount. At this rate, Peter Grant calculated that it would take between 30 and 46 years for selection to increase the beak size and body weight of the medium ground finch to match those of the large ground finch. But these changes occurred in response to an extraordinary environmental crisis. The data suggest that selection doesn't generally push consistently in one direction. Instead, in the Galápagos evolutionary change seems to go in fits and starts, moving traits one way and then another. So, let's suppose that a net change in beak size like the one that occurred during 1977 occurs only once in every century. Then it would take about 2000 years to transform the medium ground finch into the large ground finch, still a very rapid process.

Similar rates of evolutionary change have been observed elsewhere when species invade new habitats. For example, about 100,000 years ago a population of elk (called red deer in Great Britain) colonized and then became isolated on the island of Jersey, off the French coast, presumably by rising sea levels. By the time the island was reconnected with the mainland approximately 6000 years later, the red deer had shrunk to about the size of a large dog. University of Michigan paleontologist Philip Gingerich compiled data on the rate of evolutionary change in 104 cases in which species

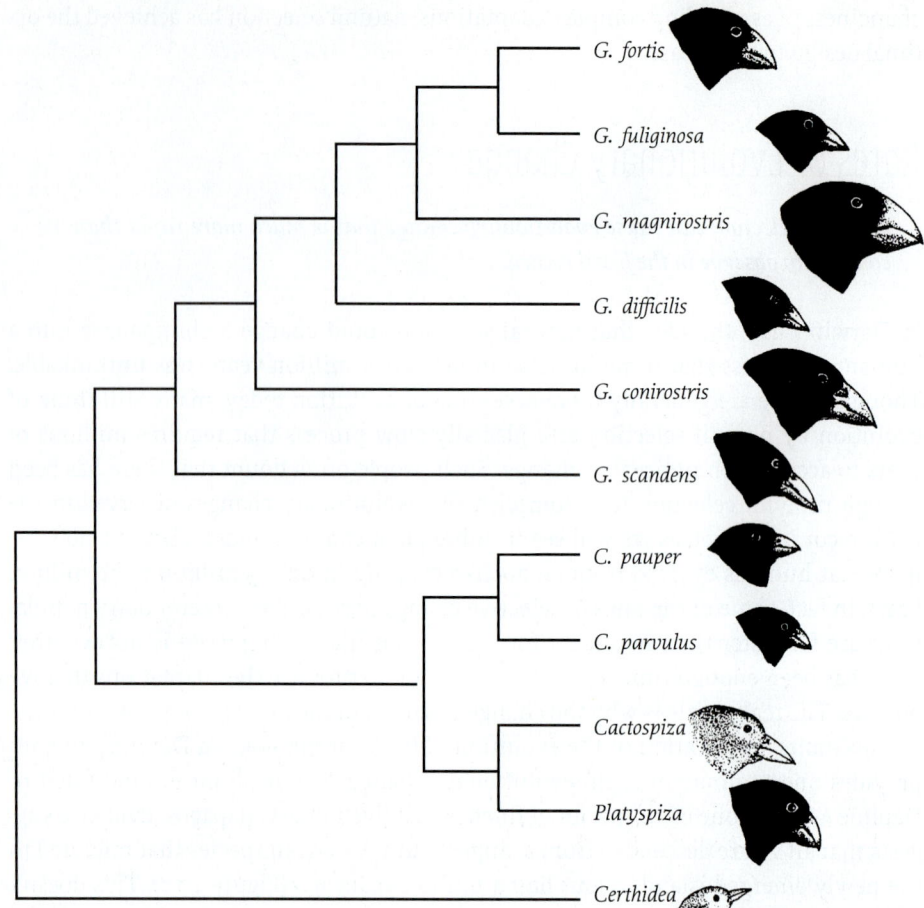

FIGURE 1.17

The relationship among various species of Darwin's finches can be traced by analyzing their protein polymorphisms. Species that are closely linked in the phylogenetic tree are more similar to one another genetically than to other species, because they share a more recent common ancestor. The tree does not include 3 of the 14 species of Darwin's finches.

FIGURE 1.18

The large ground finch (*Geospiza magnirostris*) has a beak that is nearly 20% deeper than the beak of its close relative, the medium ground finch (*G. fortis*). At the rate of evolution observed during the drought of 1977, Peter Grant calculated that selection could transform the medium ground finch into the large ground finch in less than 46 years. (Photograph courtesy of Peter and Rosemary Grant.)

invaded new habitats. These rates ranged from a low of zero (that is, no change) to a high of about 22% per year, with an average of about 0.1% per year.

The changes the Grants observed in the medium ground finch are relatively simple: birds and their beaks just got bigger. More complex changes usually take longer to evolve, but two kinds of evidence suggest that selection can produce big changes in remarkably short periods of time.

One line of evidence comes from artificial selection. Humans have performed selection on domesticated plants and animals for thousands of years, and for most of this period this selection was not deliberate. For example, modern cattle are descended from the auroch, an extinct creature something like a large antelope. People presumably kept the docile aurochs that were easy to manage and ate the rest. By inadvertently selecting for docility, a life-threatening trait in the wild, Stone Age farmers transformed the fierce auroch into the placid Hereford. There are many other familiar examples. All domesticated dogs, for instance, are believed to be descendants of wolves. Scientists are not sure when dogs were domesticated, but 10,000 years ago is a good guess, which means that in a few thousand generations, selection changed wolves into Pekinese, beagles, greyhounds, and St. Bernards. In reality, most of these breeds were created fairly recently, and most were the products of directed breeding. Darwin's favorite example of artificial selection was the domestication of pigeons. In the 19th century, pigeon breeding was a popular hobby, especially among working people who competed to produce showy birds

(Figure 1.19). They created a menagerie of wildly different forms, all descended from the rather plain-looking rock pigeon. Darwin pointed out that these breeds are so different that if they were discovered in nature, biologists would surely classify them as members of different species. Yet they were produced by artificial selection within a few hundred years.

A second line of evidence comes from theoretical studies of the evolution of complex characters. Dan-E Nilsson and Susanne Pleger of Lund University, in Sweden, have built a mathematical model of the evolution of the eye in an aquatic organism. They start with a population of organisms, each with a simple eyespot, a flat patch of light-sensitive tissue sandwiched between a transparent protective layer and a layer of dark pigment. They then consider the effect of every possible small (1%) deformation of the shape of the eyespot on the resolving power of the eye. They determine which 1% change has the greatest positive effect on the eye's resolving power, and

(a)

(b)

(c)

(d)

FIGURE 1.19
In Darwin's day, pigeon fanciers created many new breeds of pigeons, including (a) pouters, (b) fantails, and (c) carriers, from (d) the common rock pigeon.

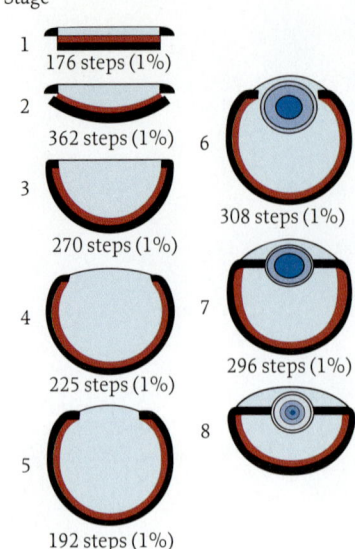

FIGURE 1.20
A computer simulation of the evolution of the eye generates this sequence of forms. In the first step, the eye is a flat patch of light-sensitive cells that lies between a transparent protective layer and a layer of dark pigment. Eventually the eyespot deepens, the lens curves outward, and the radius lengthens. This process involves about 1800 changes of 1%.

then repeat the process again and again, deforming the new structure by 1% in every possible way at each step. The results are shown in Figure 1.20. After 538 changes of 1%, a simple concave eye cup evolves; after 1033 changes of 1%, crude pinhole eyes emerge; after 1225 changes of 1%, an eye with an elliptical lens is created; and after 1829 steps the process finally comes to a halt because no small changes increase resolving power. The end result is an eye with a spherical gradient lens just like those in fish and other aquatic organisms. As Nilsson and Pleger point out, 1829 changes of 1% add up to a substantial amount of change. For instance, 1829 changes of 1% would lengthen a 10-cm (4-in.) human finger to a length of 8000 km (5500 miles)—about the distance from Los Angeles to New York and back. Nonetheless, making very conservative assumptions about the strength of selection, Nilsson and Pleger calculate that this would take only about 364,000 generations. For organisms with short generations, the complete structure of the eye can evolve from a simple eyespot in less than a million years, a brief moment in evolutionary time.

By comparison, most changes observed in the fossil record are much slower. Human brain size has roughly doubled in the last 2 million years—a rate of change of 0.00005% per year. This is 10,000 times slower than the rate of change that the Grants observed in the Galápagos. Moreover, such slow rates of change typify what can be observed from the fossil record. As we will see, however, the fossil record is incomplete. It is quite likely that some evolutionary changes in the past were rapid, but the sparseness of the fossil record prevents us from detecting them.

Darwin's Difficulties Explaining Variation

Darwin's *The Origin of Species* was a best seller during his day, but his proposal that new species and other major evolutionary changes arise by the accumulation of small variations through natural selection was not widely embraced. Most educated people accepted the idea that new species arise through the transformation of existing species, and many scientists accepted the idea that natural selection is the most important cause of organic change (although by the turn of the century even this consensus broke down, particularly in the United States). But only a minority endorsed Darwin's view that major changes occur through the accumulation of small variations.

 Darwin couldn't convince his contemporaries that evolution occurred through the accumulation of small variations because he couldn't explain how variation is maintained.

Darwin's critics raised a telling objection to his theory: the actions of blending inheritance (see below) and selection would both inevitably deplete variation in populations and make it impossible for natural selection to continue. These were potent objections that Darwin was unable to resolve in his lifetime because he and his contemporaries did not yet understand the mechanics of inheritance.

Everyone could readily observe that many of the characteristics of offspring are an average of the characteristics of their parents. Most people, including Darwin, believed this phenomena to be caused by the action of **blending inheritance,** a model of inheritance that assumes the mother and father each contribute a hereditary substance that mixes, or "blends," to determine the characteristics of the offspring. Shortly after publication of *The Origin of Species*, a Scottish engineer named Fleeming

Jenkin published a paper in which he clearly showed that with blending inheritance there could be little or no variation available for selection to act on. The following example shows why Jenkin's argument was so compelling. Suppose a population of one species of Darwin's finches displays two forms, tall and short. Further suppose that a biologist controls mating so that every mating is between a tall individual and a short individual. Then, with blending inheritance all of the offspring will be the same intermediate height, and their offspring will be the same height as they are. All of the variation for height in the population will disappear in a single generation. With random mating, the same thing will occur, though it will take longer. If inheritance were purely a matter of blending parental traits, then Jenkin would have been right about its effect on variation. However, as we will see in Chapter 3, genetics accounts for the fact that offspring are intermediate between their parents and does not assume any kind of blending.

Another problem arose because selection works by removing variants from populations. For example, if finches with small beaks are more likely to die than finches with larger beaks over the course of many generations, eventually all that will be left are birds with large beaks. There will be no variation for beak size, and Darwin's second postulate holds that without variation there can be no evolution by natural selection. For example, suppose the environment changes so that individuals with small beaks are less likely to die than those with large beaks are. The average beak size in the population will not decrease because there are no small-beaked individuals. Natural selection destroys the variation required to create adaptations.

Even worse, as Jenkin also pointed out, there was no explanation of how a population might evolve beyond its original range of variation. The cumulative evolution of complex adaptations requires that populations move far outside their original range of variation. Selection can cull away some characters from a population, but how can it lead to new types not present in the original population? This apparent contradiction was a serious impediment to explaining the logic of evolution. How could elephants, moles, bats, and whales all descend from an ancient shrewlike insectivore unless there is some mechanism for creating new variants not present at the beginning? For that matter, how could all the different breeds of dogs have descended from their one common ancestor, the wolf (Figure 1.21)?

Remember that Darwin and his contemporaries knew there were two kinds of variation: continuous and discontinuous. Because Darwin believed complex adaptations could arise only through the accumulation of small variations, he thought discontinuous variants were unimportant. However, many biologists thought the discontinuous variants, called "sports" by 19th-century animal breeders, were the key to evolution because they solved the problem of the blending effect. The following hypothetical example illustrates why. Suppose that a population of green birds has entered a new environment in which red birds would be better adapted. Some of Darwin's critics believed that any new variant that was slightly more red would have only a small advantage and would be rapidly swamped by blending. In contrast, an all-red bird would have a large enough selective advantage to overcome the effects of blending, and could increase its frequency in the population.

Darwin's letters show that these criticisms worried him greatly, and although he tried a variety of counterarguments, he never found a satisfactory one. The solution to these problems required an understanding of genetics, which was not available for another half century. As we will see, it was not until well into the 20th century that geneticists came to understand how variation is maintained, and Darwin's theory of evolution was generally accepted.

(a)

(b)

FIGURE 1.21

(a) The wolf is the ancestor of all domestic dogs, including (b) this poodle and St. Bernard. These transformations were accomplished in several thousand generations of artificial selection.

Further Reading

Browne, J. 1996. *Charles Darwin: Voyaging, A Biography.* Princeton University Press, Princeton, N.J.
Dawkins, R. 1996. *The Blind Watchmaker.* W. W. Norton, New York.
Dennett, D. 1995. *Darwin's Dangerous Idea.* Touchstone, New York.
Ridley, M. 1996. *Evolution.* 2d ed. Blackwell Scientific, Oxford.
Weiner, J. 1994. *The Beak of the Finch.* Alfred A. Knopf, New York.

Study Questions

1. It is sometimes observed that offspring do not resemble their parents for some character, even though the character varies in the population. Suppose this were the case for beak depth in the medium ground finch.
   (a) What would the plot of offspring beak depth against parental beak depth look like?
   (b) Plot the mean depth in the population among (i) adults before a drought, (ii) adults after a year of drought, and (iii) offspring.
2. Many species of animals are cannibalistic. This practice certainly reduces the ability of the species to survive. Is it possible that cannibalism could arise by natural selection? If so, with what adaptive advantage?
3. Some insects mimic dung. Ever since Darwin, biologists have explained this as a form of camouflage: selection favors individuals who most resemble dung because they are less likely to be eaten. The late Harvard paleontologist Stephen Jay Gould objected to this explanation. He argued that while selection could perfect such mimicry once it evolved, it could not cause the resemblance to arise in the first place. "Can there be any edge," Gould asks, "to looking 5% like a turd?" (quoted on p. 81 in R. Dawkins, 1996; *The Blind Watchmaker,* W. W. Norton). Can you think of a reason why looking 5% like a turd would be better than not looking at all like a turd?
4. In the late 1800s an American biologist named Hermon Bumpus collected a large number of sparrows that had been killed in a severe ice storm. He found that birds whose wings were about average length were rare among the dead birds. What kind of selection is this? What effect would this episode of selection have on the mean wing length in the population?

CHAPTER 5

Introduction to the Primates

Two Reasons to Study Primates
    Primates Are Our Closest Relatives
    Primates Are a Diverse Order
Features That Define the Primates
Primate Biogeography
A Taxonomy of Living Primates
    The Prosimians
    The Anthropoids
Primate Conservation

Two Reasons to Study Primates

The chapters in Part Two focus on the behavior of living nonhuman primates. Studies of nonhuman primates help us understand human evolution for two complementary but distinct reasons. First, closely related species tend to be similar morphologically. As we saw in Chapter 4, this similarity is due to the fact that closely related species retain and share traits acquired through descent from a common ancestor. Thus, **viviparity** (bearing live young) and lactation are traits that all placental and marsupial mammals share, and these traits distinguish mammals from other taxa, such as reptiles. The existence of such similarities means that studies of living primates often give us more insight about the behavior of our ancestors than do studies of other organisms. This approach is called "reasoning by homology." The second reason we study primates is based on the idea that natural selection leads to similar organisms in similar environments. By assessing the patterns of diversity in behavior and morphology of organisms in relation to their environments, we can see how evolution shapes adaptation in response to different selective pressures. This approach is called "reasoning by analogy."

Primates Are Our Closest Relatives

 The fact that humans and other primates share many characteristics means other primates provide valuable insights about early humans.

We humans are more closely related to nonhuman primates than we are to any other animal species. The anatomical similarities among monkeys, apes, and humans led the Swedish naturalist Carl Linnaeus to place humans in the order Primates in the first scientific taxonomy, *Systema Naturae*, published in 1758. Later, naturalists such as Georges Cuvier and Johann Blumenbach placed humans in their own order because of our distinctive mental capacities and upright posture. However, in *The Descent of Man*, Charles Darwin firmly advocated reinstating humans in the order Primates; he cited biologist T. H. Huxley's essay enumerating the many anatomical similarities between us and apes, and suggested that "if man had not been his own classifier, he would never have thought of founding a separate order for his own reception." Modern systematics unambiguously confirms that humans are more closely related to other primates than to any other living creatures.

Because we are closely related to other primates, we share with them many aspects of morphology, physiology, and development. For example, like other primates, we have well-developed visual abilities and grasping hands and feet. We share certain features of our life history with other primates, including an extended period of juvenile development, and primates as a whole have larger brains in relation to body size than the members of most other taxonomic groups do. Homologies between humans and other primates also extend to behavior, since the physiological and cognitive structures that underlie human behavior are more similar to those of other primates than to members of other taxonomic groups. The existence of this extensive array of homologous traits, the product of the common evolutionary history of the primates, means that nonhuman primates provide useful models for understanding the evolutionary roots of human morphology and for unraveling the origins of human nature.

Primates Are a Diverse Order

 Diversity within the primate order helps us to understand how natural selection shapes behavior.

During the last 30 years, hundreds of researchers from several academic disciplines have spent thousands of hours observing many different species of nonhuman primates in the wild, in captive colonies, and in laboratories. All primate species have evolved adaptations that enable them to meet the basic challenges of life, such as finding food, avoiding predators, obtaining mates, rearing young, and coping with competitors. At the same time, there is great morphological, ecological, and behavioral diversity among species within the primate order. For example, they range in size from the tiny mouse lemur, which weighs less than 30 g ($<$ 1 oz) to male gorillas weighing 160 kg (352 lb). Some species live in dense tropical forests, while others are at home in open woodlands and savannas. Some subsist almost entirely on leaves, while others rely on an omnivorous diet that includes fruit, leaves, flowers, seeds, gum, nectar, insects, and small animal prey. Some species are solitary, and others are highly gregarious. Some are active at night **(nocturnal)**; others are active during daylight hours **(diurnal).** One primate, the fat-tailed dwarf lemur, enters a torpid state and sleeps for six months each year. Some species actively defend territories from incursions by other members of their own species **(conspecifics)**; others do not. In some species, only females provide care of their young, while in others males participate actively in this process.

This variety is inherently interesting in and of itself. Researchers who study primates are motivated by absorption in the lives of their subjects to endure the hardships of fieldwork, the frustrations of attempting to obtain a share of ever-shrinking research funds, and the puzzlement of family and friends who wonder why they have chosen such an odd occupation. However, evidence of diversity among closely related organisms living under somewhat different ecological and social conditions also helps researchers to understand how evolution shapes behavior. Animals that are closely related to one another phylogenetically tend to be very similar in morphology, physiology, life history, and behavior. Thus, differences observed among closely related species are likely to represent adaptive responses to specific ecological conditions. At the same time, similarities among more distantly related creatures living under similar ecological conditions are likely to be the product of convergence.

This approach, sometimes called the comparative method, has become an important form of analysis as researchers attempt to explain the patterns of variation in morphology and behavior observed in nature. The same principles have been borrowed to reconstruct the behavior of extinct hominids, early members of the human lineage. Since behavior leaves virtually no trace in the fossil record, the comparative method provides one of our only objective means of testing hypotheses about the lives of our hominid ancestors. For example, the observation that there are substantial differences in male and female body size, a phenomenon called **sexual dimorphism,** in species that form nonmonogamous groups suggests that highly dimorphic hominids in the past were not monogamous. We will see in Part Three how the data and theories about behavior produced by primatologists have played an important role in reshaping our ideas about human origins.

Features That Define the Primates

The primate order is generally defined by a number of shared, derived characters, but not all primates share all of these traits.

The animals pictured in Figure 5.1 are all members of the primate order. These animals are similar in many ways: they are covered with a thick coat of hair, they have four limbs, and they have five fingers on each hand. However, these ancestral features are shared with all mammals. Beyond these ancestral features, it is hard to see what this group of animals has in common that makes them distinct from other mammals. What distinguishes a ring-tailed lemur from a mongoose or raccoon? What features link the langur and the aye-aye?

In fact, primates are a rather nondescript mammalian order that cannot be unambiguously characterized by a single derived feature shared by all members. However, in his extensive treatise on primate evolution, University of Zurich biologist Robert Martin defines the primate order in terms of the derived features listed in Table 5.1.

FIGURE 5.1

All of these animals are primates (a = aye-aye, b = ring-tailed lemur, c = langur, d = howler, e = gelada baboon). Primates are a diverse order and do not possess a suite of traits that unambiguously distinguish them from other animals.

The first three traits in Table 5.1 are related to the flexible movement of hands and feet. Primates can grasp with their hands and feet (Figure 5.2a), and most monkeys and apes can oppose their thumb and forefinger in a precision grip (Figure 5.2b). The flat nails, distinct from the claws of many animals, and the tactile pads on the tips of primate fingers and toes further enhance their dexterity (Figure 5.2c). These traits enable primates to use their hands and feet differently than most other animals. Primates are able to grasp fruit, squirming insects, and other small items in their hands and feet, and they can grip branches with their fingers and toes. During groom-

FEATURES THAT DEFINE THE PRIMATES

TABLE 5.1 Definition of the primate order. See the text for more complete descriptions of these features.

1. The big toe on the foot is **opposable,** and hands are **prehensile.** This means that primates can use their feet and hands for grasping. The opposable big toe has been lost in humans.
2. There are flat nails on the hands and feet in most species, instead of claws, and there are sensitive tactile pads with "fingerprints" on fingers and toes.
3. Locomotion is **hindlimb dominated,** meaning the hindlimbs do most of the work, and the center of gravity is nearer the hindlimbs than the forelimbs.
4. There is an unspecialized **olfactory** (smelling) apparatus that is reduced in diurnal primates.
5. The visual sense is highly developed. The eyes are large and moved forward in the head, providing stereoscopic vision.
6. Females have small litters, and gestation and juvenile periods are longer than in other mammals of similar size.
7. The brain is large compared with the brains of similarly sized mammals, and it has a number of unique anatomical features.
8. The **molars** are relatively unspecialized, and there is a maximum of two **incisors,** one **canine,** three **premolars,** and three molars on each half of the upper and lower jaw.
9. There are a number of other subtle anatomical characteristics that are useful to systematists but are hard to interpret functionally.

(a)

(b)

(c)

FIGURE 5.2

(a) Primates have grasping feet, which they use to climb, cling to branches, hold food, and scratch themselves.
(b) Primates can oppose the thumb and forefinger in a precision grip, a feature that enables them to hold food in one hand while they are feeding, to pick small ticks and bits of debris from their hair while grooming, and (in some species) to use tools. (c) Most primates have flat nails on their hands and sensitive tactile pads on the tips of their fingers.

FIGURE 5.3

In most primates, the eyes are moved forward in the head. The field of vision of the two eyes overlaps, creating binocular stereoscopic vision. (Photograph courtesy of Carola Borries.)

ing sessions, they delicately part their partner's hair and use their thumb and forefinger to remove small bits of debris from the skin.

The next two traits in Table 5.1 are related to a shift in emphasis among the sense organs. Primates are generally characterized by a greater reliance on visual stimuli and a reduced reliance on olfactory stimuli than other mammals are. Many primate species can perceive color, and their eyes are set forward in the head, providing them with binocular stereoscopic vision (Figure 5.3). **Binocular vision** means that the fields of vision of the two eyes overlap so that both eyes perceive the same image. **Stereoscopic vision** means that each eye sends a signal of the visual image to both hemispheres in the brain to create an image with depth. These trends are not uniformly expressed within the primate order, as olfactory cues play a more important role in the lives of prosimian primates than of anthropoid primates. As we will explain shortly, the **prosimian** primates include the lorises and lemurs, while the **anthropoid** primates include the monkeys and apes.

The next two features in Table 5.1 result from the distinctive life history of primates. As a group, primates have longer pregnancies, mature at later ages, live longer, and have larger brains than other animals of similar body size do. These features reflect a progressive trend toward increased dependence on complex behavior, learning, and behavioral flexibility within the primate order. As the noted primatologist Alison Jolly points out, "If there is an essence of being a primate, it is the progressive evolution of intelligence as a way of life." As we will see in the chapters that follow, these traits have a profound impact on the mating and parenting strategies of males and females and the patterns of social interaction among members of the order Primates.

The eighth feature in Table 5.1 concerns primate dentition. Teeth play a very important role in the lives of primates and in our understanding of their evolution. The utility of teeth to primates themselves is straightforward: teeth are necessary for processing food and are also used as weapons in conflicts with other animals. Teeth are also useful features for those who study living and fossil primates. Primatologists sometimes rely on tooth wear to gauge the age of individuals, and they use features of the teeth to assess the phylogenetic relationships among species. As we will see, paleontologists often rely on teeth, which are hard and preserve well, to identify the phylogenetic relationships of extinct creatures and to make inferences about their developmental patterns, their dietary preferences, and their social structure. Box 5.1 describes primate dentition in greater detail.

Although these traits are generally characteristic of primates, you should keep two points in mind. First, none of them makes primates unique. Dolphins, for example, have large brains and extended periods of juvenile dependence, and their social behavior may be just as complicated and flexible as that of any nonhuman primate (Figure 5.4). Second, not every primate possesses all of these traits. Humans have lost the grasping big toe that characterizes other primates, and some prosimians have claws on some of their fingers and toes.

FIGURE 5.4

A high degree of intelligence characterizes some animals besides primates. Dolphins, for example, have very large brains in relation to their body size, and behavior that is quite complex.

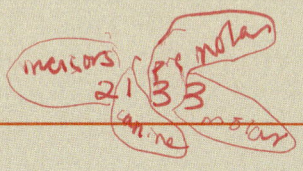

Box 5.1
What's in a Tooth?

To appreciate the basic features of primate dentition, you can consult Figure 5.12, or you can simply look in a mirror since your teeth are much like those of other primates. Teeth are rooted in the jaw. The jaw holds four different kinds of teeth; in order they are first the incisors at the front, then the canines, premolars, and the molars in the rear. To understand what each kind of teeth does, imagine yourself eating a sandwich. You bite into the sandwich with your incisors and canines and use your front teeth to detach a piece. The incisors are relatively small, peg-shaped teeth. The canines are dagger-shaped. The upper canines are usually considerably longer than the other teeth, and the upper canine is sharpened on the lower premolar. The incisors and canines are mainly involved in getting food into the mouth and preparing food for further processing by the molars and premolars. The molars and premolars have broad surfaces, covered by a series of enamel bumps, or **cusps,** connected by crests or ridges. The molars and premolars are mainly used to crush, shred, and chew food before it is swallowed. In Chapter 6 we will show how the size and shape of the teeth are related to the types of foods primates eat.

Although all primates have the same kinds of teeth, species vary in how many of each kind of tooth they have. For convenience, these combinations are expressed in a standard format called the **dental formula,** which is commonly written in the following form:

$$\frac{2.1.3.3}{2.1.3.3}$$

The numerals separated by periods tell us how many of each of the four types of teeth a particular species has (or had) on one side of its jaw. Left to right, the four types are given for the front of the mouth (incisors) to the rear (molars). The top line represents the teeth on one side of the upper jaw **(maxilla),** and the bottom line represents the teeth on the corresponding side of the lower jaw **(mandible).** Hence this species—which happens to be the common ancestor of all primates—had two incisors, one canine, three premolars, and three molars each on one side of the upper and lower jaws. Usually, but not always, the formula is the same for both upper and lower jaws. Like most other parts of the body, our dentition is **bilaterally symmetric,** which means that the left side is identical to the right side. The ancestral pattern shown here has been modified in various primate taxa, as the total number of teeth has been reduced (Table 5.2).

The dental formulas among living primates vary. Prosimians have the most variable dentition. While the lorises, pottos, galagos, and a number of lemurids have retained the primitive dental formula, other groups have lost incisors, canines, or premolars. Tarsiers have lost one incisor on the mandible but retained two on the maxilla. Dentition is generally less variable among the anthropoid primates than among the prosimians. All of the New World monkeys, except the marmosets and tamarins, have retained the primitive dental formula; the marmosets and tamarins have lost one molar. The Old World monkeys, apes, and humans have reduced the number of premolars from 3 to 2.

TABLE 5.2 Primates vary in the numbers of each type of tooth that they have. The dental formulas here give the number of incisors, canines, premolars, and molars on each side of the upper and lower jaw. For example, lorises have two incisors, one canine, three premolars, and three molars on each side of their upper and lower jaw.

Primate Group	Dental Formula
Prosimians	
Lorises, pottos, and galagos	$\frac{2.1.3.3}{2.1.3.3}$
Dwarf lemurs, mouse lemurs, and true lemurs	$\frac{2.1.3.3}{2.1.3.3}$
Indris	$\frac{2.1.2.3}{2.0.2.3}$
Aye-aye	$\frac{1.0.1.3}{1.0.0.3}$
Tarsiers	$\frac{2.1.3.3}{1.1.3.3}$
New World monkeys	
Most species	$\frac{2.1.3.3}{2.1.3.3}$
Marmosets and tamarins	$\frac{2.1.3.2}{2.1.3.2}$
All Old World monkeys, apes, and humans	$\frac{2.1.2.3}{2.1.2.3}$

Primate Biogeography

Primates are generally restricted to tropical regions of the world.

The continents of Asia, Africa, and South America and the islands that lie near their coasts are home to most of the world's primates (Figure 5.5). A few species remain in Mexico and Central America. Primates were once found in southern Europe, but no natural populations survive there now. There are no natural populations of primates in Australia or Antarctica, and none occupied these continents in the past.

Primates are mainly found in tropical regions, where the fluctuations in temperature from day to night greatly exceed fluctuations in temperature over the course of the year. In the tropics, the distribution of resources that primates rely on for subsistence is more strongly affected by seasonal changes in rainfall than by seasonal changes in temperature. Some primate species extend their ranges into temperate areas of Africa and Asia, where they manage to cope with substantial seasonal fluctuations in environmental conditions.

Within their ranges, primates occupy an extremely diverse set of habitats that includes all types of tropical forests, savanna woodlands, mangrove swamps, grasslands, high-altitude plateaus, and deserts. The vast majority of primates, however, are found in forested areas, where they travel, feed, socialize, and sleep in a largely arboreal world.

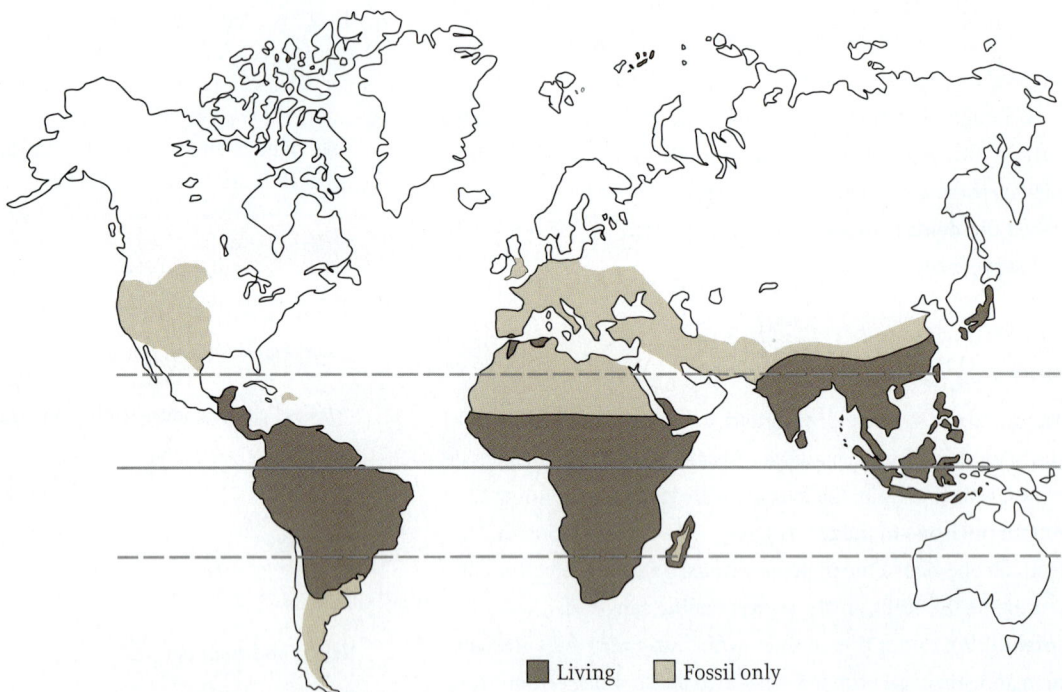

FIGURE 5.5

The distribution of living and fossil primates is shown here. Primates are now found in Central America, South America, Africa, and Asia. They are mainly found in tropical regions of the world. Primates were formerly found in southern Europe and northern Africa. There have never been indigenous populations of primates in Australia or Antarctica.

A Taxonomy of Living Primates

 The primates are divided into two groups, the prosimians and the anthropoids.

Scientists classify primates into two suborders, the Prosimii and Anthropoidea (Table 5.3). Many of the primates included in the suborder Prosimii are nocturnal, and like some of the earliest primates that lived 50 mya, they have many adaptations to living in darkness, including a well-developed sense of smell, large eyes, and independently movable ears. By contrast, monkeys and apes, which make up the suborder Anthropoidea, evolved adaptations more suited to a diurnal lifestyle early in their evolutionary history. In the Anthropoidea, the traits related to increased complexity of behavior are most fully developed.

The classification of the primates into prosimians and anthropoids does not strictly reflect the patterns of genetic relationships among the animals in the suborders. Tarsiers are included in the prosimians because like the lorises and lemurs, they are nocturnal creatures that have retained many ancestral characters. However, both genetic and morphological data suggest that tarsiers are more closely related to monkeys and apes than to prosimians. Thus, a purely cladistic classification would place tarsiers in the same **infraorder** (the taxonomic level immediately below suborder) as the monkeys. In fact, many primate taxonomists advocate a taxonomy in which the lemurs and lorises are classified together as **strepsirhines** and the rest of the primates are classified together as the **haplorhines.** The more traditional division into prosimians and anthropoids is an example of evolutionary systematics in which overall similarity and relatedness are used to classify species.

THE PROSIMIANS

 The prosimians are divided into three infraorders: Lemuriformes, Lorisiformes, and Tarsiiformes.

The infraorder Lemuriformes includes lemurs, which are found only on Madagascar and the Comoro Islands, off the southeastern coast of Africa. These islands have been separated from Africa for 120 million years. The primitive prosimians that reached Madagascar evolved in total isolation from primates elsewhere in the world, as well as from many of the predators and competitors that primates confront in other places. Faced with a diverse set of available ecological niches, the lemurs underwent a spectacular adaptive radiation. When humans first colonized Madagascar about 2000 years ago, there were approximately 44 species of lemurs, ranging in size from mouse lemurs that weigh less than 30 g ($<$ 1 oz) to lemurs that were as large as gorillas (200 kg, 440 lb). Within the next few centuries, all of the larger lemur species became extinct, probably the victims of human hunters. The extant lemurs are mainly small or medium-sized arboreal residents of forested areas; they travel quadrupedally or by jumping in an upright posture from one tree to another, a form of locomotion known as **vertical clinging and leaping.** Activity patterns of lemurs are quite variable: about half are primarily diurnal, others are nocturnal, and some are active during both day and night. One of the most interesting aspects of lemur behavior is that females routinely dominate males. In most lemur species, females are able to supplant males from desirable feeding sites, and in some lemur species females regularly defeat males in aggressive en-

TABLE 5.3 A taxonomy of the living primates. (From R. Martin, 1992; Classification of Primates, pp. 20–21 in S. Jones, R. Martin, and D. Pilbeam, 1994, *The Cambridge Encyclopedia of Human Evolution,* Cambridge University Press, Cambridge.)

Suborder	Infraorder	Superfamily	Family	Subfamily
Prosimii (prosimians)	Lemuriformes	Lemuroidea (lemurs)	Cheirogaleidae (dwarf and mouse lemurs)	
			Lemuridae	Lemurinae (true lemurs)
				Lepilemurinae (sportive lemurs)
			Indridae (indris)	
			Daubentoniidae (aye-ayes)	
	Lorisiformes	Lorisoidea (loris group)	Lorisidae	Lorisinae (lorises)
				Galaginae (galagos)
	Tarsiiformes	Tarsioidea	Tarsiidae (tarsiers)	
Anthropoidea (anthropoids)	Platyrrhini	Ceboidea (New World monkeys)	Cebidae	Cebinae (e.g., capuchins)
				Aotinae (e.g., owl monkeys)
				Atelinae (e.g., spider monkeys)
				Alouattinae (howling monkeys)
				Pithecinae (e.g., sakis)
				Callimiconinae (Goeldi's monkeys)
			Callitrichidae (marmosets and tamarins)	
	Catarrhini	Cercopithecoidea (Old World monkeys)	Cercopithecidae	Cercopithecinae (e.g., macaques and vervets)
				Colobinae (e.g., langurs)
		Hominoidea (apes and humans)	Hylobatidae	Hylobatinae (gibbons and siamangs)
			Pongidae	Ponginae (great apes)
			Hominidae	Homininae (humans)

Orangutan Conservation

[*Author's note:* For orangutans to survive into and through the next millennium, immediate action must be taken. This piece documents the orangutan's plight and evaluates some of the efforts being made to save orangutans in the wild.]

Since prehistoric times, hunting by humans has resulted in greatly reduced populations of orangutans. Today, some indigenous cultures still hunt orangutans for food but a far greater problem is habitat loss due to logging and conversion to agriculture. Orangutans do not cope well with the effects of habitat exploitation by humans. First, they are old-growth specialists, which makes them sensitive to forest disturbance such as selective logging, and they disappear altogether from heavily logged forest or cleared land. Orangutans simply cannot survive in deforested areas because they require such large home ranges and depend on a large diversity of tree and liana species. Second, their dependence on forest and reluctance to travel across open areas makes the fragmentation effects of logging and development more serious for them than for virtually any other forest species. Finally, logging, and especially conversion, tend to be concentrated in habitats such as alluvial flats that are preferred by orangutans.

As little as four decades ago, Borneo and Sumatra were almost completely covered by tropical forests. Deforestation has changed the landscape dramatically, leaving only isolated pockets of protected habitats suitable for orangutans. Initially, logging was restricted to accessible lowland areas, but excessive logging has pushed the frontier into hitherto inaccessible swampy and steep, hilly terrain in remote regions. This logging has traditionally been highly selective, focusing on only a small number of valuable export species. However, selectively logged areas have almost invariably been converted into agricultural areas by burgeoning local populations or, increasingly, as part of an integrated program of transmigration and large-scale agricultural plantations. As a result of these developments, [it is estimated that] orangutan numbers in Borneo and Sumatra had fallen to approximately 25,000 individuals, by the late 1990s, less than 8% of those estimated almost a century earlier.

Activities that damage or destroy natural forests have increased with the growing population and economies of Southeast Asia. The main orangutan concentration in Sumatra is within the Leuser Ecosystem in the northern part of the island. Recent estimates indicate that logging and clearing for agricultural plantations in Leuser have led to a 45% decline in orangutan numbers over the past seven years alone. The current wave of illegal logging, brought about by the anarchy following the fall of the Suharto regime, is increasing the pace of this already steep decline. Matters would have been even worse if the massive forest fires of 1997 and 1998 had not missed the orangutan's range in Sumatra. Nonetheless, the total estimated number in Sumatra has now fallen well below 10,000.

Borneo was less fortunate, however. A recent survey suggests that massive fires alone caused a 33% loss of the remaining population of orangutans on that island. We can only guess what the current wave of illegal logging and mining is doing to their numbers on this island, but people have now invaded the Tanjung Puting National Park, one of the orangutan's main refuges in Borneo, while Kutai National Park, a former stronghold, has all but disappeared. With the exception of orangutans within the Gunung Palung reserve, the remaining animals are scattered over numerous forest fragments, almost all of which may prove too small to retain viable populations.

The relationship between orangutans and their habitats is one of interdependence. Orangutans act as seed dispersers and predators that help maintain the species diversity of the forests they inhabit. For this reason, the presence of orangutans is a good indicator of the biological diversity of Southeast Asian rainforests. If orangutans are present at normal densities, then the area is likely also to contain at least five other species of primates, at least five species of hornbills, at least 50 different fruit-tree species, and 15 liana species. Thus, orangutans are an excellent "umbrella species" for rainforest conservation. This species' requirements with regard to area and habitat are wide enough that if orangutans were made a focus of protective management, the biodiversity of species

within its range would also automatically be preserved. Furthermore, it is extremely important to conserve as much habitat as possible to maintain variability within and between orangutan populations so that we can better understand inter-island differences and study cultural variation, which may be rapidly eroding.

National and international conservation organizations have responded to threats to orangutans by setting up rehabilitation centers for confiscated animals and by establishing protected areas. The development of rehabilitation centers that reintroduce confiscated and formerly captive orangutans into the wild has been one attempt to preserve viable populations. Initially, the task of such centers was to squelch the trade in pets, and they released animals into forests with existing orangutan populations. In addition, these centers have become the basis for ecotourism programs that encourage the viewing of wild orangutans. Ecotourism can produce revenues to continue and improve conservation efforts and provide local people with economic incentives not to destroy the extensive forest tracts that sustain orangutans and sympatric species. More recent attempts at reintroduction have focused on returning animals, increasingly displaced by forest conversion or driven out by forest fires, into suitable but currently unoccupied forests.

Various national parks are in place in regions of Borneo and Sumatra where orangutans range. However, the protection of these parks is inadequate, and the recent lawlessness has made it harder to defend the orangutan's key habitats. In the absence of effective enforcement of existing conservation policies, including the tight regulation of logging, orangutan populations are currently slipping from endangered to critically endangered. This problem is compounded where one country contains species found nowhere else: Indonesia contains over 90% of the world's wild orangutans.

Any further losses may spell the end of a variety of endangered species, including the wild orangutan. In order to prevent these extinctions, uncontrolled and unplanned logging and clearing must be brought to a halt. Only in well-protected old-growth forests does the wild orangutan have a realistic hope of long-term survival. And only through continued field studies on wild orangutans can we hope to understand those as yet unsolved puzzles we outlined in the introduction and develop realistic models of hominid behavioral ecology, as well as the evolution of many unique human behaviors.

SOURCE: From pp. 201–208 in R. A. Delgado and C. P. van Schaik, 2000, The behavioral ecology and conservation of the orangutan *(Pongo pygmaeus)*: a tale of two islands, *Evolutionary Anthropology*, 9:201–218.

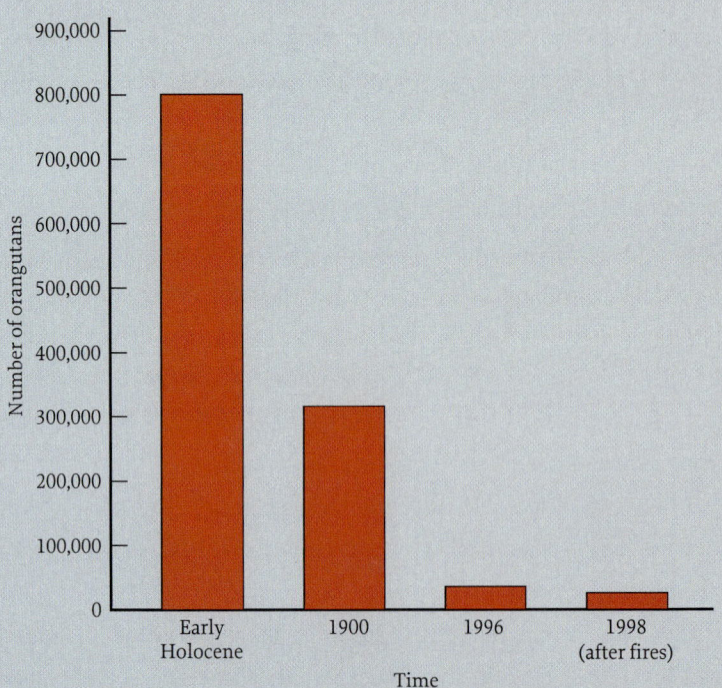

Total estimated orangutan populations since the early Holocene. Data are taken from H. D. Rijksen and E. Meijaard, 1999, *Our Vanishing Relative*. Dordrecht: Kluwer Academic Publishers.

A TAXONOMY OF LIVING PRIMATES

counters. Alison Jolly, one of the first observers of free-ranging lemurs, noted, "At any time a female may casually supplant any male or irritably cuff him over the nose and take a tamarind pod from his hand." While such behavior may seem unremarkable in our own liberated times, female dominance is very rare in other primate species.

The infraorder Lorisiformes comprises small, nocturnal, arboreal residents of the forests of Africa and Asia (Figure 5.6). These animals include two subfamilies with different types of locomotion and activity patterns. Galagos are active and agile, leaping through the trees and running quickly along branches. The lorises move with ponderous deliberation, and their wrists and ankles have a specialized network of blood vessels that allows them to remain immobile for long periods of time. These traits may be adaptations that help them avoid detection by predators. Traveling alone, the lorisiforms generally feed on fruit, gum, and insect prey. The lorisiforms leave their dependent offspring in nests built in the hollows of trees or hidden in masses of tangled vegetation. During the day, females sleep, nurse their young, and groom, sometimes in the company of mature offspring or familiar neighbors.

The infraorder Tarsiiformes includes tarsiers, which are enigmatic primates who live in the rain forests of Borneo, Sulawesi, and the Philippines (Figure 5.7). Like many prosimians, they are small, nocturnal, and arboreal and move by vertical clinging and leaping. Some tarsiers live in monogamous family groups, but many have more than one breeding female. Female tarsiers give birth to infants who weigh 25% of their own weight; mothers leave their bulky infants behind in safe hiding places when they forage for insects. Tarsiers are unique among primates because they are the only primate that relies exclusively on animal matter, feeding on insects and small vertebrate prey.

FIGURE 5.6
Galagos are small, arboreal, nocturnal animals who can leap great distances. They are mainly solitary, though residents of neighboring territories sometimes rest together during the day.

FIGURE 5.7
Tarsiers are small, insectivorous primates who live in Asia. Some tarsiers form monogamous pairs.

THE ANTHROPOIDS

The suborder Anthropoidea contains the infraorders Platyrrhini and Catarrhini.

The two infraorders Platyrrhini and Catarrhini are commonly referred to as the New World monkeys and the Old World monkeys and apes, respectively, because platyrrhine monkeys are found in South and Central America, while catarrhine monkeys and apes are found in Africa and Asia. This geographic dichotomy breaks down with humans, however—we are catarrhine primates, but we are spread over the globe.

Platyrrhini, or New World monkeys, is divided into two families: Callitrichidae and Cebidae.

Although the New World monkeys encompass considerable diversity in size, diet, and social organization, they do share some basic features. All but those in one genus *(Aotus)* are diurnal, all live in forested areas, and all are mainly arboreal. Most New World monkeys are quadrupedal, moving along the tops of branches and jumping between adjacent trees. However, some species within the family Cebidae can suspend themselves by their hands, feet, or tail and can move by swinging by their arms beneath branches.

The family Callitrichidae is composed of the marmosets and tamarins. These species share several morphological features that distinguish them from other anthropoid primate species: they are extremely small, the largest weighing less than 1 kg (2.2 lb), they have claws instead of nails, they have only two molars while all other monkeys have three, and they frequently give birth to twins and sometimes triplets (Figure 5.8). Marmosets and tamarins are also notable for their domestic arrangements: most species seem to be monogamous, and some may be polyandrous. **Polyandry,** which occurs when two or more males simultaneously form pair-bonds with a single female, is extremely uncommon among mammalian species. Marmoset and tamarin mothers receive a considerable amount of assistance in caring for their young from their mates and older offspring.

The cebid monkeys are generally larger than the marmosets and tamarins, ranging in size from the 600-g (21-oz) squirrel monkey to the 9.5-kg (21-lb) muriqui (Figure 5.9). Although many people think that all monkeys can swing by their tails, prehensile tails are actually restricted to only the largest species of the Cebidae.

The family Cebidae is divided into six subfamilies which encompass considerable diversity in social organization, feeding behavior, and ecology. The subfamily Alouattinae is composed of several species of howlers, named for the long-distance roars they give in intergroup interactions. Howlers live in small one-male or multimale groups, defend their home ranges, and feed mainly on leaves. The subfamily Atelinae includes spider monkeys and woolly monkeys (Figure 5.9a). These species subsist mainly on fruit and leaves, and live in multimale, multifemale groups of 15 to 25. Spider monkeys, which rely heavily on ripe fruit, typically break up into small parties for feeding (Figure 5.9b). The subfamily Cebinae includes capuchins and squirrel monkeys. Capuchins are best known to the public as the clever creatures dressed in red caps and jackets that retrieve coins for organ grinders (Figure 5.9c). To primatologists, capuchins are notable in part because they have very large brains in relation to their body size (see Chapter 9). Squirrel monkeys and capuchins live in multimale, multifemale groups of 10 to 50 individuals and forage for fruit, leaves, and insects (Figure 5.9d). The subfamily Aotinae includes the diurnal titi monkey and the only nocturnal anthropoid primate, the owl monkey. The subfamily Callimiconinae is composed of just one species, Goeldi's monkey. Aotinae and Callimiconinae monkeys are small-bodied fruit eaters and live in monogamous groups. Finally, Pithecinae is a subfamily composed of the uakaris and sakis. Uakaris have little hair around their faces, making them look like wizened old men.

FIGURE 5.8

Marmosets are small-bodied South American monkeys who form monogamous or polyandrous social groups. Males and older offspring participate actively in the care of infants.

The infraorder Catarrhini contains the monkeys and apes of the Old World and humans.

As a group, the catarrhine primates share a number of anatomical and behavioral features that distinguish them from the New World primates. For example, most Old World monkeys and apes have narrow nostrils that face downward, while New World monkeys have round nostrils. Old World monkeys have two premolars on each side of the upper and lower jaws, while New World monkeys have three. Most Old World primates are larger than most New World species, and Old World monkeys and apes occupy a wider range of habitats than New World species do.

(a)

(c)

(b)

(d)

FIGURE 5.9

Portraits of some cebid monkeys: (a) Muriquis, or woolly spider monkeys, are large-bodied and arboreal. They are extremely peaceful creatures, rarely fighting or competing over access to resources. (b) Spider monkeys rely heavily on ripe fruit and travel in small parties. They have prehensile tails that they can use much like an extra hand or foot. (c) Capuchin monkeys have larger brains in relation to their body sizes than any of the other nonhuman primates. (d) Squirrel monkeys form large multimale, multifemale groups. In the mating season, males gain weight and become "fatted," and then compete actively for access to receptive females. (Photographs courtesy of: a, Sue Boinski; c, Susan Perry; d, Carlão Limeira.)

The catarrhine primates are divided into two superfamilies, the Cercopithecoidea (Old World monkeys) and Hominoidea (apes and humans). Cercopithecoidea contains one extant (still-living) family, which is further divided into two subfamilies of monkeys, Cercopithecinae and Colobinae.

The superfamily Cercopithecoidea encompasses great diversity in social organization, ecological specializations, and biogeography.

Colobine monkeys are found in the forests of Africa and Asia and are collectively perhaps the most elegant of the primates (Figure 5.10). They have slender bodies, long legs, long tails, and often beautifully colored coats. The black-and-white colobus, for example, has a white ring around its black face, a striking white cape on its black back, and a bushy white tail that flies out behind as it leaps from tree to tree. Colobines are mainly leaf and seed eaters, and most species spend the majority of their time in trees. They have complex stomachs, almost like the chambered ones of cows, that allow them to maintain bacterial colonies that facilitate the digestion of cellulose. Colobines are most often found in groups composed of one adult male and

FIGURE 5.10

(a) African colobines, like this black-and-white colobus monkey, are arboreal and feed mainly on leaves. These animals are sometimes hunted for their spectacular coats. (b) Hanuman langurs are native to India and have been the subject of extensive study during the last three decades. In some areas, Hanuman langurs form one-male, multifemale groups, and males engage in fierce fights over membership in bisexual groups. In these groups, infanticide often follows when a new male takes over the group. (b, Photograph courtesy of Carola Borries.)

(a)   (b)

a number of adult females. As in many other vertebrate taxa, the replacement of resident males in one-male groups is often accompanied by lethal attacks on infants by new males. Infanticide under such circumstances is believed to be favored by selection because it improves the relative reproductive success of infanticidal males. This issue is discussed more fully in Chapter 7.

Most cercopithecine monkeys are found in Africa, though one successful genus (*Macaca*) is widely distributed through Asia and part of Europe (Figure 5.11). The cercopithecines are more variable in size and diet than the colobines are. The social behavior, reproductive behavior, life history, and ecology of a number of cercopithecine species (particularly baboons, macaques, and vervets) have been extensively studied. Cercopithecines typically live in medium or large bisexual (multimale, multifemale) groups. Females typically remain in their **natal groups** (the groups into which they are born) throughout their lives and establish close and enduring relationships with their maternal kin, while males leave their natal groups and join new groups when they reach sexual maturity.

The superfamily Hominoidea includes three families of apes: Hylobatidae (gibbons), Pongidae (orangutans, gorillas, and chimpanzees), and Hominidae (humans).

The hominoids are different from the cercopithecoids in a number of ways. The most readily observed difference between apes and monkeys is that apes lack tails. But there are many other more subtle differences between apes and monkeys. For example, the apes share some derived traits, including broader noses, broader palates, and larger brains, and they retain some primitive traits, such as relatively unspecial-

A TAXONOMY OF LIVING PRIMATES

FIGURE 5.11

Some representative cercopithecines: (a) Bonnet macaques are one of several species of macaques that are found throughout Asia and North Africa. Like other macaques, bonnet macaques form multimale, multifemale groups, and females spend their entire lives in their natal (birth) groups. (b) Vervet monkeys are found throughout Africa. Like macaques and baboons, females live among their mothers, daughters, and other maternal kin. Males transfer to nonnatal groups when they reach maturity. Vervets defend their ranges against incursions by members of other groups. (c) Blue monkeys live in one-male, multifemale groups. However, during the mating season, one or more unfamiliar males may join bisexual groups and mate with females. (Photographs courtesy of: a, Kathy West; c, Marina Cords.)

ized molars. In Old World monkeys the prominent anterior and posterior cusps are arranged to form two parallel ridges. In apes, the five cusps on the lower molars are arranged to form a side-turned Y-shaped pattern of ridges (Figure 5.12).

The family Hylobatidae includes lesser apes, while the larger-bodied Pongidae (orangutans, gorillas, and chimpanzees) encompasses great apes. Humans are traditionally placed in their own family, the Hominidae, but many taxonomists believe

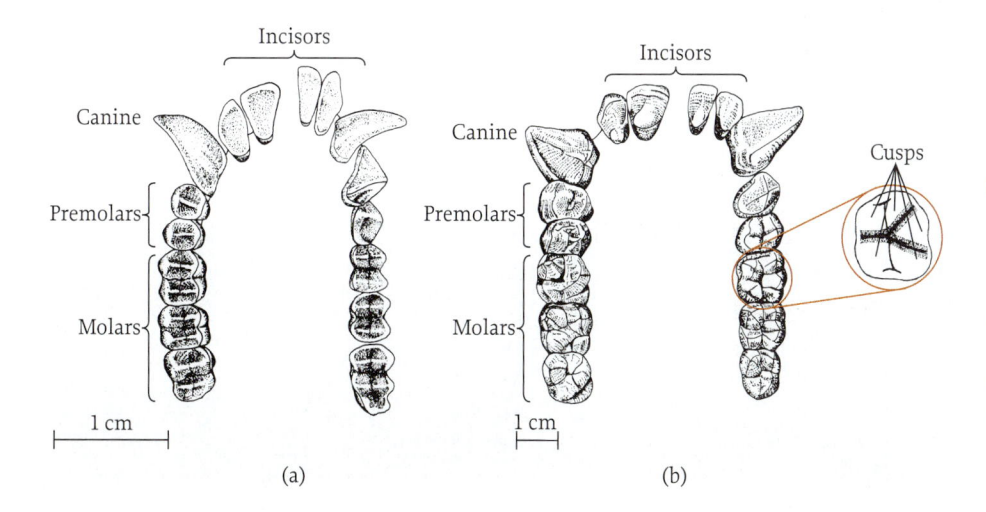

FIGURE 5.12

The upper jaw (left) and lower jaw (right) are shown here for (a) a male colobine and (b) a male gorilla. In Old World monkeys the prominent anterior and posterior cusps of the lower molars form two parallel ridges. In apes the five cusps of the lower molar form a Y-shaped pattern.

that humans belong with the other large-bodied apes, in the Pongidae. Gibbons, siamangs, and orangutans are found in Asia, while chimpanzees, bonobos, and gorillas are restricted to Africa.

The lesser apes (Figure 5.13) are slightly built creatures with extremely long arms in relation to their body size. Gibbons and siamangs are strictly arboreal, and they use their long arms to perform spectacular acrobatic feats, moving through the canopy with grace, speed, and agility. Gibbons and siamangs are the only true **brachiators** among the primates, which means they propel themselves by their arms alone and are in free flight between handholds. (To picture this, think about swinging on monkey bars in your elementary school playground.) All of the lesser apes live in monogamous family groups, vigorously defend their **home ranges** (the areas they occupy), and feed on fruit, leaves, flowers, and insects. Siamang males play an active role in caring for young, frequently carrying them during the day, while male gibbons are less attentive fathers. In territorial displays, mated pairs of siamangs perform coordinated vocal duets that can be heard over long distances.

Orangutans, now found only on the Southeast Asian islands of Sumatra and Borneo, are among the largest and most solitary species of primates (Figure 5.14). Orangutans have been extensively studied by Biruté Galdikas in Tanjung Puting, Borneo, for more than 20 years. Ongoing studies of orangutans are also being conducted at Cabang Panti in Borneo, and at Ketambe and Suaq Balimbing in Sumatra. Orangutans feed primarily on fruit, but also eat some leaves and bark. Adult females associate mainly with their own infants and immature offspring and do not often meet or interact with other orangutans. Adult males spend the majority of their time alone. A single adult male may defend a home range that encompasses the home ranges of several adult females, while other males wander over larger areas and mate opportunistically with receptive females. When resident males encounter these nomads, fierce and noisy encounters may take place.

Gorillas, the largest of the apes, existed in splendid isolation from Western science until the middle of the 19th century (Figure 5.15). Today, our knowledge of the behavior and ecology of gorillas is based mainly on detailed long-term studies of one subspecies, the mountain gorilla, at the Karisoke Research Center in Rwanda founded by the late Dian Fossey. Mountain gorillas live in small groups that contain one or two adult males and a number of adult females and their young. Each day, mountain goril-

FIGURE 5.13

(a) Gibbons live in monogamous groups and actively defend their territories against intruders. They have extremely long arms, which they use to propel themselves from one branch to another as they swing hand over hand through the canopy, a form of locomotion called brachiation. Gibbons and siamangs (b) are confined to the tropical forests of Asia. Like other residents of tropical forests, their survival is threatened by the rapid destruction of tropical forests. (Photographs courtesy of John Mitani.)

FIGURE 5.14

(a) Orangutans are large, ponderous, and mostly solitary creatures. Male orangutans often descend to the ground to travel, while lighter females often move through the tree canopy. (b) Today, orangutans are found only on the islands of Borneo and Sumatra, in tropical forests like this one.

A TAXONOMY OF LIVING PRIMATES

(a)   (b)

FIGURE 5.15

(a) Gorillas are the largest of the primates. Mountain gorillas usually live in one-male, multifemale groups, but some groups contain more than one adult male.
(b) Most behavioral information about gorillas comes from observations of mountain gorillas who live in the Virunga Mountains of Central Africa pictured here. The harsh montane habitat may influence the nature of social organization and social behavior in these animals, and the behavior of gorillas living at lower elevations may differ. (Photographs courtesy of John Mitani.)

las ingest great quantities of various herbs, vines, shrubs, and bamboo. They eat little fruit because fruiting plants are scarce in their mountainous habitat. Adult male mountain gorillas, called **silverbacks** because the hair on their backs and shoulders turns a striking silver-gray when they mature, play a central role in the structure and cohesion of their social groups. Males sometimes remain in their natal groups to breed, but most males leave their natal groups and acquire females by drawing them away from other males during intergroup encounters. The silverback largely determines the timing of group activity and the direction of travel. As data from newly established field studies of lowland gorilla populations become available, we may have reason to revise some elements of this view of gorilla social organization. For example, lowland gorillas seem to eat substantial amounts of fruit, spend more of their time in trees, and form larger and less cohesive social groups than mountain gorillas do.

As humankind's closest living relatives, chimpanzees (Figure 5.16a) have played a uniquely important role in the study of human evolution. Whether reasoning by homology or by analogy, researchers have found observations about chimpanzees to be important bases for hypotheses about the behavior of early hominids.

Detailed knowledge of chimpanzee behavior and ecology comes from a number of long-term studies conducted at sites across Africa. In the 1960s, Jane Goodall began her well-known study of chimpanzees at the Gombe Stream National Park on the shores of Lake Tanganyika in Tanzania (Figure 5.16b). About the same time, a second study was initiated by Toshisada Nishida at a site in the Mahale Mountains not far from Gombe. These studies are now moving into their fourth decade. Other important study sites have been established at Boussou, Guinea, in the Taï Forest of the Ivory Coast, and at two sites in the Kibale Forest of Uganda, Kanyawara and Ngogo.

Bonobos (Figure 5.16c), another member of the genus *Pan*, live in inaccessible places and are much less well-studied than common chimpanzees. Important field studies on bonobos have been conducted at two sites in the Democratic Republic of the Congo, Wamba and Lomako. Field studies of bonobos have been disrupted by civil conflicts that have ravaged Central Africa over the last decade.

Chimpanzees and bonobos form large multimale, multifemale communities. These communities differ from the social groups formed by most other species of primates in two important ways. First, female chimpanzees usually disperse from their natal groups when they reach sexual maturity, while males remain in their

(a) (b) (c)

FIGURE 5.16
(a) Chimpanzees live in multimale, multifemale social groups. In this species, males form the core of the social group and remain in their natal groups for life. Many researchers believe that chimpanzees are our closest living relatives. (b) Like other apes, chimpanzees are found mainly in forests like this area on the shores of Lake Tanganyika in Tanzania. However, chimpanzees sometimes range into more open areas as well.
(c) Bonobos are members of the same genus as chimpanzees and are similar in many ways. Bonobos are sometimes called pygmy chimpanzees, but this is a misnomer because bonobos and chimpanzees are about the same size. This infant bonobo is sitting in a patch of terrestial herbaceous vegetation, one of the staples of the bonobo's diet. (b, c, Photographs courtesy of John Mitani.)

natal groups throughout their lives. Second, the members of chimpanzee communities are rarely found together in a unified group. Instead, they split up into smaller parties that vary in size and composition from day to day. In chimpanzees, the strongest social bonds among adults are formed among males, while bonobo females form stronger bonds with one another and with their adult sons than males do. Chimpanzees modify natural objects for use as tools in the wild. At several sites, chimpanzees strip twigs and poke them into termite mounds and ant nests to extract insects, a much-prized delicacy. In the Taï Forest, chimpanzees crack hard-shelled nuts using one stone as a hammer and a heavy, flat stone or a protruding root as an anvil. At Gombe, chimpanzees wad leaves in their mouths, and then dip these "sponges" into crevices to soak up water. New data also reveal tool use by wild orangutans, but chimpanzee tool use is more diverse and better-studied.

Primate Conservation

Many species of primates are endangered by 1) habitat destruction, 2) hunting, or 3) live capture for trade and export.

Sadly, no introduction to the primate order would be complete without acknowledging that the prospects for the continued survival of many primate species are grim. Ninety-six primate species are now considered to be endangered or critically endangered, and are in real danger of extinction. Already, primate conservation biologists believe that one subspecies, Miss Waldron's red colobus, has become

PRIMATE CONSERVATION

extinct. The populations of some of the most gravely endangered species, such as the mountain gorilla and the golden lion tamarin, number in the hundreds. In Africa and Asia, one-third to one-half of all primate species are endangered, while in some parts of South America an even larger fraction of species are at risk (Figure 5.17). At least two thirds of the lemur species in Madagascar are in immediate danger of extinction.

As arboreal residents of the tropics, most primate populations are directly affected by the rapid and widespread destruction of the world's forests. Primate ecologists Colin Chapman of the University of Florida and Carlos Peres of the University of East Anglia recently reviewed the conservation status of the world's primate populations. Their analysis is quite sobering. Between 1980 and 1995, approximately 10% of the forests in Africa and Latin America were lost, while 6% of the forests of Asia disappeared (Figure 5.18). Countries that house primates are losing about 125,000 km^2 of forest *every year*, an area about the size of the state of Mississippi.

The destruction of tropical forests is the product of economic and demographic pressures acting on governments and local residents. Many developing countries have huge foreign debts that must be repaid. The need to raise funds to pay off these debts generates intense pressure for timber harvesting and more intensive agricultural activity. Each year, 5 to 6 million hectares of forest is logged, seriously disrupting the lives of the animals that live in them. (A hectare is a square, 100 m on a side, or about 2.5 acres.)

FIGURE 5.17

Many populations of primates are in danger of extinction. This is a free-ranging member of a subspecies of squirrel monkeys that may number only 200 in the wild.

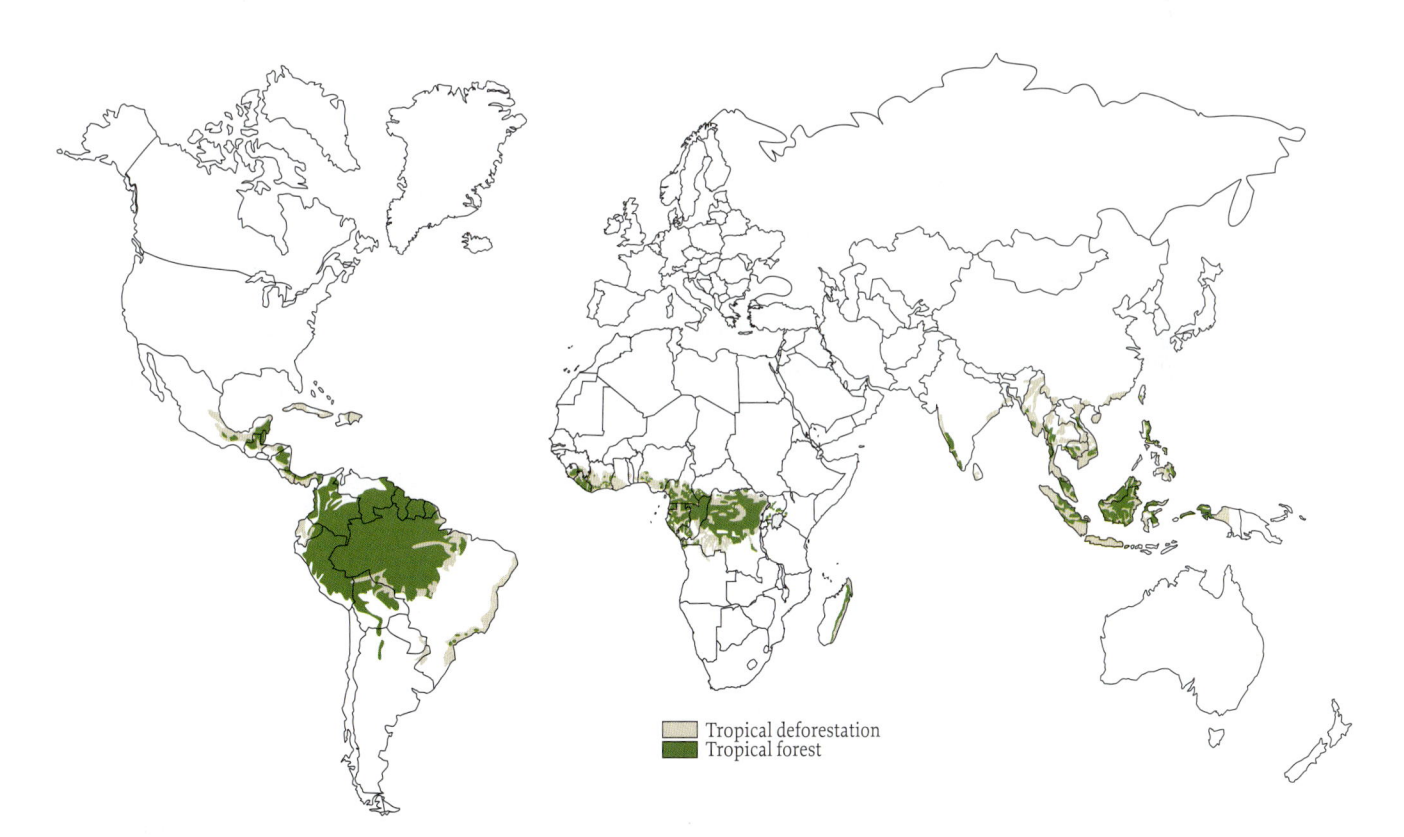

FIGURE 5.18

This map shows the major locations of tropical forest and the areas that have become deforested (90% of the canopy cover has been lost). Deforestation is a major threat to primates because many primate species live in tropical forests.

FIGURE 5.19

The destruction of tropical forests is often related to population pressures. Here, data on deforestation and human population growth are plotted for some of the countries that harbor free-ranging primate populations. Countries that have high rates of population growth have the highest rates of deforestation.

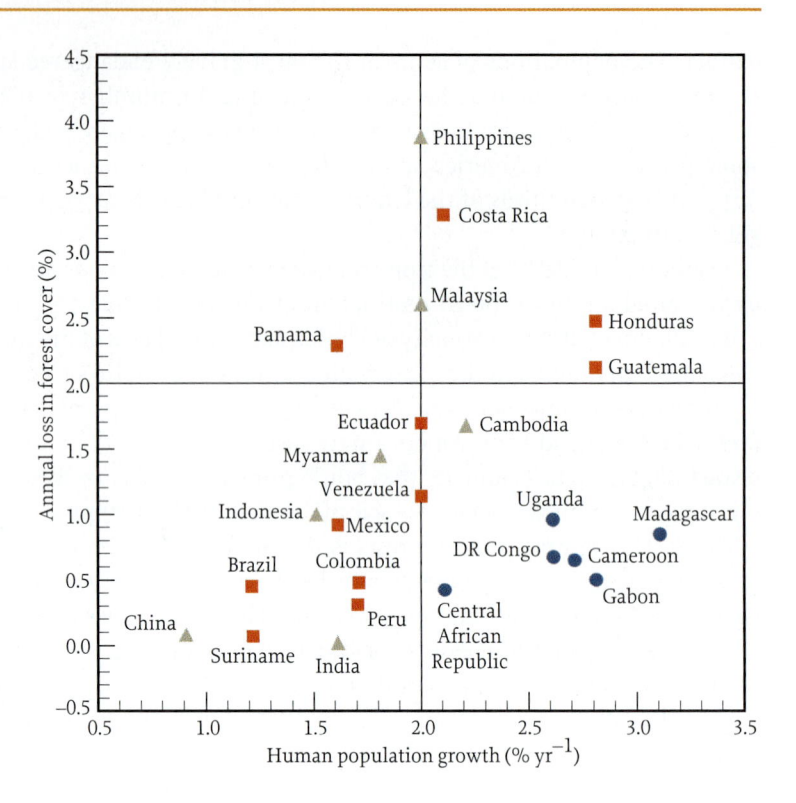

Forests are also cleared for agricultural activities. Rapid increases in the population of underdeveloped countries in the tropics have created intense demand for additional agricultural land (Figure 5.19). For example, in West Africa, Asia, and South America, vast expanses of forests have been cleared to accommodate the demands of subsistence farmers trying to feed their families, as well as the needs of large-scale agricultural projects (Figure 5.20). In Central and South America, massive areas have been cleared for large cattle ranches.

In the last two decades, a new threat to the forests of the world has emerged—wildfire. Major fires have destroyed massive tracts of forest in Southeast Asia and South America. Ecologists believe that natural fires in tropical forests are relatively rare, and that these devastating fires are the product of human activity. In Indonesia, massive fires in the late 1990s left thousands of orangutans dead, reducing their numbers by nearly a third.

In many areas around the world, particularly South America and Africa, primates are also hunted for meat. Although systematic information about the impact of hunting on wild primate populations is scant, some case studies reveal troubling trends. For example, in one forest in Kenya, 1200 blue monkeys and nearly 700 baboons were killed by subsistence hunters in one year. In the Brazilian Amazon, one family of rubber tappers killed 200 woolly monkeys, 100 spider monkeys, and 80 howler monkeys during an 18-month span. In addition to subsistence hunting, there is also an active market for "bushmeat" in many urban areas.

The capture and trade of live primates has been greatly reduced since the Convention of Trade in Endangered Species of Wild Flora and Fauna (CITES) was drafted in 1973. The parties to CITES, which now number 122 countries, ban commercial trade of all endangered species and monitor the trade of those that are at risk of becoming endangered. CITES has been an effective weapon in protecting primate populations

PRIMATE CONSERVATION

FIGURE 5.20

This forest, on the border of the Lomas Barbudal National Park in Costa Rica, has just been logged. In many countries all of the land surrounding nature reserves and national parks is under cultivation, created forested islands. Although primates may be protected within these reserves, their isolation threatens their long-term survival. The elimination of surrounding forest corridors restricts the movement of migrating animals and limits the size and genetic diversity of local primate populations. (Photograph by Colin Chapman.)

around the world. The United States imported more than 100,000 primates each year before ratifying CITES, but reduced this number to approximately 13,000 a decade after signing the international agreement.

Although CITES has made a major impact, some problems persist. Thus, live capture for trade remains a major threat to certain species, particularly the great apes, whose high commercial value creates strong incentives for illegal commerce. In many communities, young primates are kept as pets (Figure 5.21). For each animal taken into captivity, many other animals are put at risk. This is because hunters cannot obtain young primates without capturing their mothers, who are usually killed in the process. In addition, many prospective pets die after capture, from injuries suffered during capture and transport or from poor housing conditions and inappropriate diets while in captivity.

Efforts to save endangered primate populations have met with some success.

Although much remains to be done, conservation efforts have significantly improved the survival prospects of a number of primate species. These efforts have helped to preserve muriquis and golden lion tamarins in Brazil and golden bamboo lemurs in Madagascar. But there is no room for complacency. Promising efforts to save orangutans in Indonesia and mountain gorillas in Rwanda have been seriously impeded by regional political struggles and armed conflict, putting the apes' habitats and their lives in serious jeopardy. A number of different strategies to conserve forest habitats and preserve animal populations are on the table. These include land-for-debt swaps in which foreign debts are forgiven in exchange for commitments to conserve natural habitats, development of ecotourism projects, and sustainable development of forest resources. But as conservationists study these solutions and try to implement them, the problems facing the world's primates become more pressing. More and more forests disappear each year, and many primates are lost, perhaps forever.

FIGURE 5.21

These two colobine monkeys were captured by poachers for the pet trade. The animals were then confiscated by the Ugandan Wildlife Authority and taken to a local zoo. (Photograph by Colin Chapman.)

Further Reading

Chapman, C. A., and C. A. Peres. 2001. Primate conservation in the new millennium: the role of scientists. *Evolutionary Anthropology* 10:16–33.

Cowlishaw, G., and R. I. M. Dunbar. 2000. *Primate Conservation Biology.* University of Chicago Press, Chicago.

Fleagle, J. G. 1998. *Primate Adaptation and Evolution.* 2d ed. Academic Press, San Diego, Calif., chap. 1.

Kramer, R. C., C. van Schaik, and J. Johnson. 1997. *Last Stand. Protected Areas and the Defense of Tropical Biodiversity.* Oxford University Press, Oxford.

Martin, R. D. 1990. *Primate Origins and Evolution: A Phylogenetic Analysis.* Princeton University Press, Princeton, N.J.

Study Questions

1. What is the difference between homology and analogy? What are the evolutionary processes that correspond to these terms?
2. Suppose that a group of extraterrestrial scientists lands on Earth and enlists your help in identifying animals. How do you help them recognize members of the primate order?
3. What kinds of habitats do most primates occupy? What are the features of this kind of environment?
4. Outline the taxonomy of the living primates to the superfamily level. Identify the geographic region the animals inhabit as well as their major features.
5. What primitive characteristics do modern prosimians retain?
6. In many ways, the superfamily Lemuroidea comprises a more diverse group than other primate superfamilies. Why is this?
7. What genera are included in the superfamily Hominoidea? Briefly describe the social organization and geographic range of each of these genera.
8. Why are contemporary primate species threatened? What are the major hazards facing them today?
9. How can we balance the needs and rights of people living in developing nations with the needs of the primates who live around them?
10. Local peoples have been living alongside monkeys and other animals in tropical forests for thousands of years. If this is the case, then why do we face the present conservation crisis? What has changed?

CHAPTER 6

Primate Ecology

THE DISTRIBUTION OF FOOD
ACTIVITY PATTERNS
RANGING BEHAVIOR
PREDATION
PRIMATE SOCIALITY
   THE DISTRIBUTION OF FEMALES
   THE DISTRIBUTION OF MALES

Much of the day-to-day life of primates is driven by two concerns: getting enough to eat and avoiding being eaten. Food is essential for growth, survival, and reproduction, and it should not be surprising that primates spend much of every day finding, processing, consuming, and digesting a wide variety of foods (Figure 6.1). At the same time, primates must always guard against predators like lions, pythons, and eagles, which hunt them by day, and leopards, which stalk them by night. As we will see in the chapters that follow, both the distribution of food and the threat of predation influence the extent of sociality among primates and shape the patterning of social interactions within and between primate groups.

In this chapter we describe the basic features of primate ecology. Later, we will draw on this information to explore the relationships among ecological factors, social organization, and primate behavior. It is important to understand the nature of these relationships because the same ecological factors are likely to have influenced the social organization and behavior of our earliest ancestors.

FIGURE 6.1

A female baboon feeds on corms in Amboseli, Kenya.

145

The Distribution of Food

Food provides energy that is essential for growth, survival, and reproduction.

Like all animals, primates need energy to maintain normal metabolic processes, to regulate essential bodily functions, and to sustain growth, development, and reproduction. The total amount of energy that an animal requires depends on four components: basal metabolism, active metabolism, growth rate, and reproductive effort.

1. **Basal metabolic rate** is the rate at which an animal expends energy to maintain life when at rest. As you can see from Figure 6.2, larger animals have higher basal metabolic rates than smaller ones do. However, large animals require relatively fewer calories *per unit* of body weight.
2. When animals become active, their energy needs rise above baseline levels. The number of additional calories required depends on how much energy the animal expends. This, in turn, depends on the size of the animal and how fast it moves. In general, to sustain a normal range of activities, an average-sized primate like a baboon or macaque requires enough energy per day to maintain a rate about twice its basal metabolic rate.
3. Growth imposes further energetic demands on organisms. Infants and juveniles, who are gaining weight and growing in stature, require more energy than would be expected on the basis of their body weight and activity levels alone.
4. In addition, for female primates the energetic costs of reproduction are also substantial. During the latter stages of their pregnancies, for example, primate females require about 25% more calories than usual, and during lactation they require about 50% more calories than usual.

A primate's diet must satisfy the animal's energy requirements, provide specific types of nutrients, and minimize exposure to dangerous toxins.

The food that primates eat provides them with energy and essential nutrients, such as amino acids and minerals, that they cannot synthesize themselves. Proteins are essential for virtually every aspect of growth and reproduction and for the regulation of many bodily functions. As we saw in Chapter 2, proteins are composed of long chains of amino acids. Primates cannot synthesize amino acids from simpler molecules, and

FIGURE 6.2
Average basal metabolism is affected by body size. The dashed line represents a direct linear relationship between body weight and basal metabolic rate. The solid line represents the actual relationship between body weight and basal metabolic rate. The fact that the curve "bends over" means that larger animals use relatively less energy per unit of body weight.

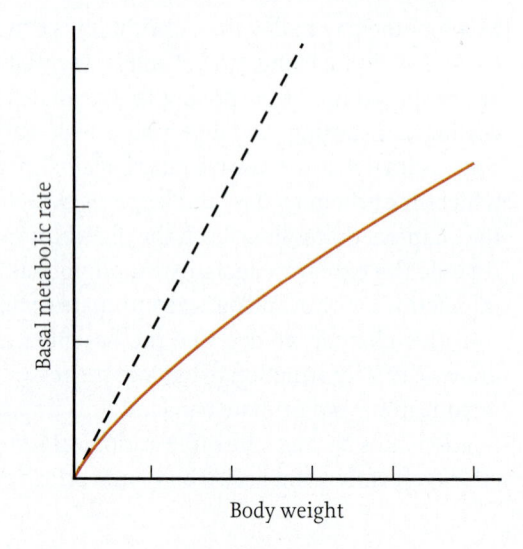

THE DISTRIBUTION OF FOOD

in order to build many essential proteins, they must ingest foods that contain sufficient amounts of a number of amino acids. Fats and oils are important sources of energy for animals and provide about twice as much energy as equivalent volumes of **carbohydrates.** Vitamins, minerals, and trace amounts of certain elements play an essential role in regulating many of the body's metabolic functions. Although they are needed in only small amounts, deficiencies of specific vitamins, minerals, or trace elements can cause significant impairment of normal body function. For example, trace amounts of the elements iron and copper are important in the synthesis of hemoglobin, while vitamin D is essential for the metabolism of calcium and phosphorus, and sodium regulates the quantity and distribution of body fluids. Primates cannot synthesize any of these compounds and must acquire them from the foods they eat. Water is the major constituent of the bodies of all animals and most plants. For survival, most animals must balance their water intake with their water loss; moderate dehydration can be debilitating, and significant dehydration can be fatal.

At the same time that primates attempt to obtain nourishment from food, they must also take care to avoid **toxins,** substances in the environment that are harmful to them. Many plants produce toxins called **secondary compounds** in order to protect themselves from being eaten. Thousands of these secondary compounds have been identified: caffeine and morphine are among the secondary compounds most familiar to us. Some secondary compounds, such as **alkaloids,** are toxic to consumers because they pass through the stomach into various types of cells, where they disrupt normal metabolic functions. Common alkaloids include capsicum (the compound that brings tears to your eyes when you eat red peppers) and chocolate. Other secondary compounds, such as tannins (the bitter-tasting compound in tea), act in the consumer's gut to reduce the digestibility of plant material. Secondary compounds are particularly common among tropical plant species and are often concentrated in mature leaves and seeds. Young leaves, fruit, and flowers tend to have lower concentrations of secondary compounds, making them relatively more palatable to primates.

Primates obtain nutrients from many different sources.

Primates obtain energy and essential nutrients from a variety of sources (Table 6.1). Carbohydrates are obtained mainly from the simple sugars in fruit, but animal prey,

TABLE 6.1 Sources of nutrients for primates. (x) indicates that the nutrient content is generally accessible only to animals that have specific digestive adaptations.

Source	Protein	Carbohydrates	Fats and Oils	Vitamins	Minerals	Water
Animals	x	(x)	x	x	x	x
Fruit		x				x
Seeds	x		x	x		
Flowers		x				x
Young leaves	x			x	x	x
Mature leaves	(x)					
Woody stems	x					
Sap		x			x	x
Gum	x	(x)			x	
Underground parts	x	x				x

Monkeys Can Be Picky Eaters

[*Authors' note:* Muriquis, or woolly spider monkeys, are large, arboreal South American monkeys. They live in large multimale, multifemale groups and are remarkably peaceful.]

Energy from fruits and flowers, and protein from leaves, are the principal constituents of the muriqui's diet. But the forest is full of plants that produce chemical and physical deterrents to avoid being eaten, and muriquis must distinguish between edible and inedible ones. Mature leaves of most woody plants contain high levels of tannins, which bind with the leaf's proteins and make it difficult to digest. The presence of tannins, as well as alkaloids and other secondary compounds, probably explains why muriquis rarely feed on any particular leaf species for any length of time. Several individuals will feed in sequence from the same tree, but each one feeds only briefly before moving to another food source. Immature leaves, however, generally contain fewer tannins, and muriquis spend more time eating larger quantities of these when they are available.

Unripe fruits tend to have more toxic compounds than the ripe fruits that muriquis prefer. When ripe fruit is plentiful in a single large tree, or food patch, the number of muriquis that feed together and the time they spend eating increases with the size of the patch. Unlike mature leaves, which must be eaten in only small quantities at a time, mature fruits can be eaten until the monkeys get bored or the food runs out. . . .

Even seeds contain toxins, or are wrapped in thick coats for protection from predators. Muriquis chew small seeds along with the fruit's flesh, but larger seeds are often dropped after the flesh has been removed. Some seeds, however, are covered with a smooth, hard coat, a design that makes them easy to swallow along with the fruit. These seeds glide through the muriquis' digestive tracts, and appear intact in their feces. . . .

A muriqui feeds on fruit. (Photograph courtesy of C. P. Nogueira.)

Muriquis serve an important ecological function in dispersing these large, smooth seeds. Because they usually leave a fruit tree before their meal has been digested, the muriquis effectively transport the seeds to another part of the forest, away from the shade and competitive environment of the mother tree. When my students collected seeds from muriqui feces and planted them elsewhere, they almost always germinated. In fact, some muriqui-dispersed seeds actually germinate faster after they have been eaten than they do without the benefit of having passed through the muriquis' digestive system.

Muriquis eat only a few species of seeds that are not surrounded by fleshy fruit. The seeds of one of these, *Mucuna* (Leguminosae), are contained in large pods which are covered with a hard shell bristling with minuscule spines that break off and implant themselves in any hand or mouth that touches them. Adult and subadult muriquis are able to open these pods and eat the large, fatty seeds, but smaller muriquis avoid *Mucuna*, perhaps because they are not strong enough to open the pods, or because the fleshy pads of their fingers are not tough enough to deflect the spines. The *Mucuna*'s spiny defenses protect it from juvenile muriquis and other potential predators, but its fatty seeds are too important for larger muriquis to resist.

Leaves from the fig-like *Cecropia* (Moraceae) are another example of a food that muriquis go to great lengths to get. *Cecropia* have co-evolved with ants, which live in their trunks and defend them by attacking and stinging any animal that tries to enter the trees to feed. To get around this defense, muriquis eat *Cecropia* leaves by suspending themselves from neighboring trees. I once witnessed an adult male attempting to reach a *Cecropia* from a small palm tree growing next to it. He rocked back and forth in the crown of the palm causing it to swing. Finally, the momentum of his weight brought the palm close enough to the *Cecropia* for him to grab a leaf before the palm swung back. He then sat in the center of the still-swaying palm tree to eat the leaf, his reward for a 10 minute effort. Biochemical analyses indicate that these *Cecropia* are especially high in protein and low in other toxic compounds, so the nutritional pay-offs appear to compensate for the time and ingenuity required to outsmart the tree's defense system.

The trade-off between nutritional gains and plant defenses may not be the only basis for muriqui food choices. Some of the secondary compounds that plants produce may also have important medicinal properties. Many indigenous human populations throughout the tropics exploit different plants and plant parts for pharmaceutical purposes, and chimpanzees and baboons are known to eat certain plants for their antiparasitic and antibacterial compounds. I suspected that the muriquis at Fazenda Montes Claros might find similarly important compounds in some of their foods, so in 1989 I began a study of the parasites in their gastrointestinal tracts. I relied on the information that could be gleaned from noninvasive examination of their feces, which [was] conducted by Dr. Michael Stuart and his students at the University of North Carolina–Asheville. Looking at over 80 fecal samples collected from 32 different monkeys, the results were astonishing. The muriquis' feces were completely devoid of parasites, and our comparative analyses of howler monkey feces at this site show that they, too, are parasite-free. Most other primates, including howler monkeys and muriquis at other forests, carry in their gastrointestinal tracts a number of parasites that are passed along in their feces, so the fact that neither of these primates at Fazenda Montes Claros [is] infected suggests that something unusual is going on.

Parasites have evolved complex ecological relationships with their hosts, and the absence of parasites in both muriquis and howlers could mean that these ecological—and evolutionary—relationships have been disrupted at this forest. Perhaps a key secondary host has become locally extinct. On the other hand, it is possible that both primate species at Fazenda Montes Claros have discovered antiparasitic agents in their diets. Muriquis and howler monkeys eat many of the same plants, and by analyzing the ones that are found only at this forest we may be able to identify bioactive plants with value, not only for monkeys, but for humans as well.

SOURCE: From pp. 53–56 in K. B. Strier, 1992, *Faces in the Forest: The Endangered Muriqui Monkeys of Brazil*, Oxford University Press, New York. Reprinted by permission of Oxford University Press.

such as insects, also provide a good source of fats and oils. **Gum,** a substance that plants produce in response to physical injury, is an important source of carbohydrates for some primates, particularly galagos, marmosets, and tamarins. Primates get most of their protein from insect prey or from young leaves. Some species have special adaptations that facilitate the breakdown of cellulose, enabling them to digest more of the protein contained in the cells of mature leaves. While seeds provide a good source of vitamins, fats, and oils, many plants package their seeds in husks or pods that shield their contents from seed predators. Many primates drink daily from streams, water holes, springs, or puddles of rainwater (Figure 6.3). Primates can also obtain water from fruit, flowers, young leaves, animal prey, and the underground storage parts (roots and tubers) of various plants. This is particularly important for arboreal animals that do not descend from the canopy of tree branches and for terrestrial animals during times of the year when surface water is scarce. Vitamins, minerals, and trace elements are obtained in small quantities from many different sources.

Although there is considerable diversity in diet among primates, some generalizations are possible:

1. All primates rely on at least one type of food high in protein and another high in carbohydrates. Prosimians generally obtain protein from insects and carbohydrates

FIGURE 6.3

Savanna baboons drink from a pool of rainwater. Most primates must drink every day.

from gum and fruit. Monkeys and apes usually obtain protein from insects or young leaves and carbohydrates from fruit.

2. Most primates rely more heavily on some types of foods than on others. Moreover, most species are fairly conservative in their dietary preferences, even though the availability of specific types of food varies over the course of the year in many habitats, and the species of food plants eaten often varies between habitats. Chimpanzees, for example, feed mainly on ripe fruit throughout their range from Tanzania to the Ivory Coast. Scientists use the terms **frugivore, folivore, insectivore,** and **gummivore** to refer to primates that rely most heavily on fruit, leaves, insects, and plant gum, respectively (Figure 6.4). Box 6.1 examines some of the adaptations in morphology among primates with different diets.

3. In general, insectivores are smaller than frugivores and frugivores are smaller than folivores (Figure 6.5). This is related to the fact that small animals have rela-

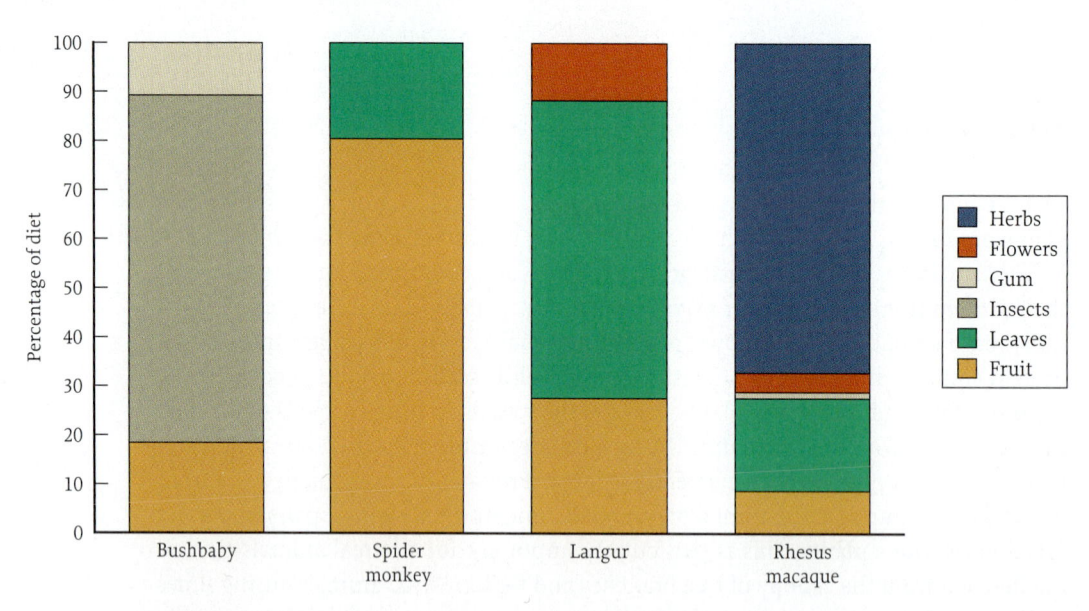

FIGURE 6.4

The diets of primates are variable. The black-handed spider monkey is one of the most dedicated frugivores in the primate order—80% of its diet consists of fruit. Langurs are among the most folivorous primate species; more than half of the purple-faced langur's diet is composed of leaves, while fruit and flowers play a much less important role. The bushbaby is primarily insectivorous, relying on insects for 70% of its food supply. Macaques have more eclectic diets, feeding on herbs, fruit, gum and sap, flowers, and leaves.

THE DISTRIBUTION OF FOOD

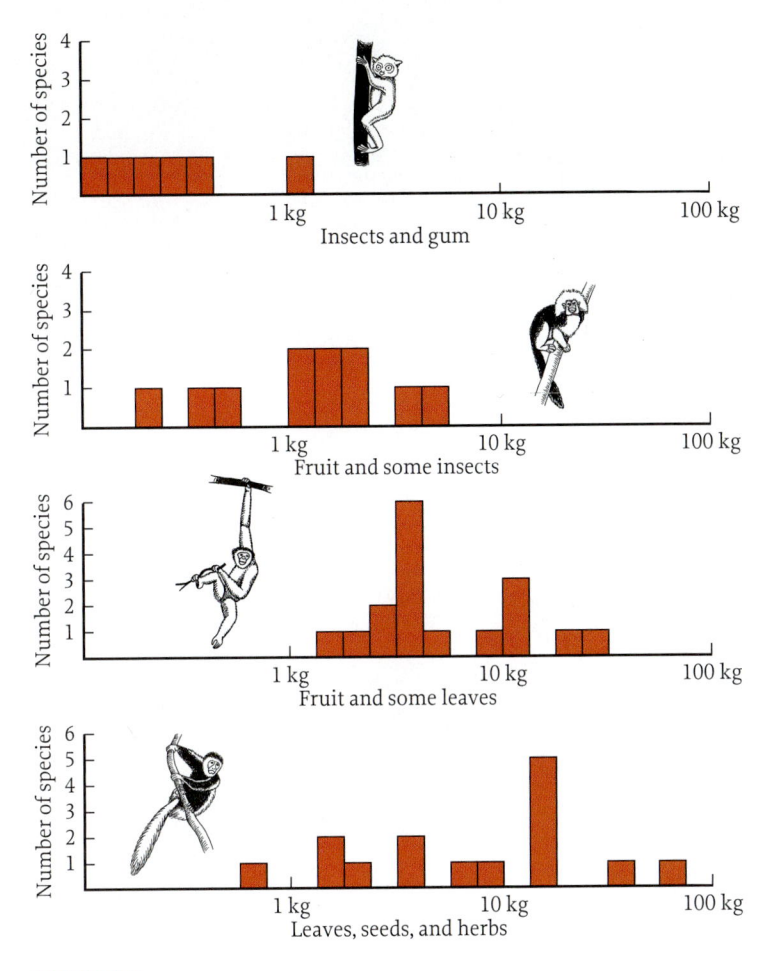

FIGURE 6.5

Body size and diet are related among primates. The smallest species mainly eat insects and gum, while the largest species eat leaves, seeds, and herbs. Fruit-eating species fall in between.

tively higher energy requirements than larger animals do, and require relatively small amounts of high-quality foods that can be processed quickly. Larger animals are less constrained by the quality of their food than by the quantity, as they can afford to process lower-quality foods more slowly.

The nature of dietary specializations and the challenge of foraging in tropical forests influence ranging patterns.

Nonhuman primates do not have the luxury of shopping in supermarkets where abundant supplies of food are concentrated in a single location and are constantly replenished. Instead, the availability of their preferred foods varies widely in space and time, making their food sources patchy and often unpredictable. Most primate species live in tropical forests. Although such forests, with their dense greenery, seem to provide abundant supplies of food for primates, appearances can be deceiving. Tropical forests contain a very large number of tree species, and individual trees of any particular species are few in number. Katherine Milton, a primatologist at the University of California at Berkeley, has conducted detailed studies of

Box 6.1
Dietary Adaptations of Primates

Primates have evolved a number of adaptations to enhance their ability to process and digest certain types of foods. Their morphological dietary specializations include specific adaptations of the teeth and gut (Figure 6.6).

TEETH Primates that rely heavily on gum tend to have large and prominent incisors, which they use to gouge holes in the bark of trees. In some prosimian species the incisors and canines are projected forward in the jaw, and are used to scrape hardened gum off the surface of branches and tree trunks. Dietary specializations are also reflected in the size and shape of the molars. Insectivores and folivores have molars with well-developed shearing crests

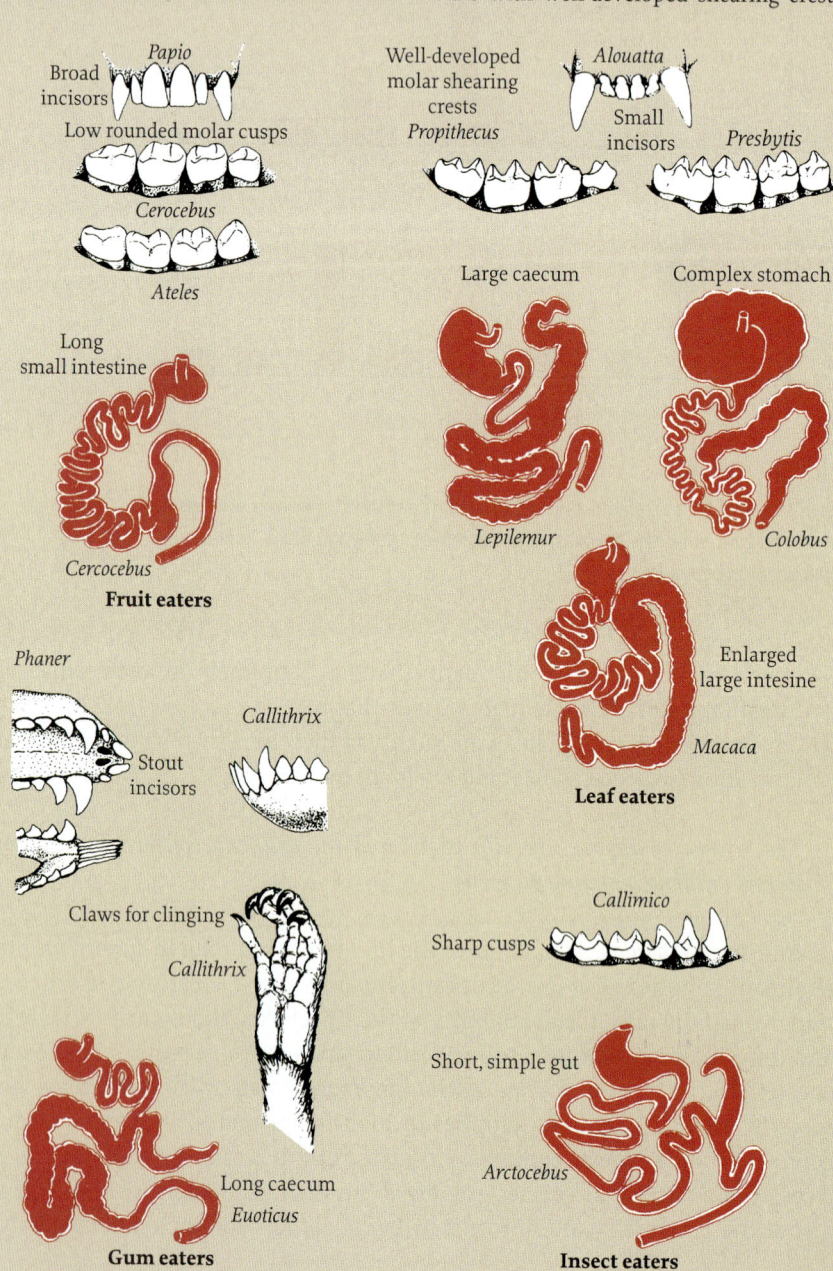

FIGURE 6.6
The dentition and digestive tracts of fruit-eating, leaf-eating, gum-eating, and insect-eating primates typically differ.

that permit them to cut their food into small pieces when they chew. Insectivores tend to have higher and more pointed cusps on their molars, which are useful for puncturing and crushing the bodies of their prey. The molars of frugivores tend to have flatter, more rounded cusps, with broad and flat areas used to crush their food. Primates that rely on hard seeds and nuts have molars with very thick enamel that can withstand the heavy chewing forces needed to process these types of food.

GUT Primates that feed principally on insects or animal prey have relatively simple digestive systems that are specialized for absorption. They generally have a simple small stomach, a small caecum (a pouch located at the upper end of the large intestine), and a small colon relative to the rest of the small intestine. Frugivores also tend to have simple digestive systems, but frugivorous species with large bodies have capacious stomachs to hold large quantities of the leaves they consume along with the fruit in their diet. Folivores have the most specialized digestive systems because they must deal with large quantities of cellulose and secondary plant compounds. Since primates cannot digest cellulose or other structural carbohydrates directly, folivores maintain colonies of microorganisms in their digestive systems that break down these substances. In some species these colonies of microorganisms are housed in an enlarged caecum, while in other species the colon is enlarged for this purpose. Colobines, for example, have an enlarged and complex stomach divided into a number of different sections where microorganisms help process cellulose.

the feeding ecology of howler monkeys in a lowland tropical forest in central Panama (Figure 6.7). In the course of her work, she carefully surveyed the composition of the forest in which the howlers live. She discovered that 65% of all tree species occur less than once per hectare. This means that potential food sources are patchily distributed in space. Moreover, Milton's studies showed there is considerable variation in the production of new leaves, flowers, and fruit over the course of the year. In the forests of central Panama, young leaves are available on a single species for an average of seven months, green and ripe fruits are available for four months, and flower parts are available for three months. On individual trees, these items are available for an even shorter period. Ripe fruit, for example, is available on individual trees for less than one month on average. In some cases food items sustain an optimal nutritional content and edibility for only a few days. Thus, foraging in a tropical forest is not an easy task.

Primates with different dietary specializations confront different foraging challenges (Figure 6.8). Plants generally produce more leaves than flowers or fruit, and bear leaves for a longer period during the year than they bear flowers and fruit. As a result, foliage is normally more abundant than fruit or flowers are at a given time during the year, and mature leaves are more abundant than young leaves. Insects and other suitable prey animals occur at even lower densities than plants. This means that folivores can generally find more food in a given area than frugivores or insectivores can. However, the high concentration of toxic secondary compounds in mature leaves complicates the foraging strategies of folivores. Some leaves must be avoided altogether, and others can be eaten only in small quantities. Nonetheless, the food supplies of folivorous species are generally more uniform and predictable in space and time than are the food supplies of

FIGURE 6.7

This is the forest of Barro Colorado Island, where howler monkeys have been studied for many years.

FIGURE 6.8

(a) Some primates feed mainly on leaves, though many leaves contain toxic secondary plant compounds. These primates are red colobus monkeys in the Kibale Forest of Uganda. (b) Some primates include a variety of insects and other animal prey in their diet. This capuchin monkey in Costa Rica is feeding on a wasp's nest. (c) Mountain gorillas are mainly vegetarians. They consume vast quantities of plant material, like this fibrous stem. (d) A vervet monkey feeds on grass stems. (e) Although many primates feed mainly on one type of food, such as leaves or fruit, none relies exclusively on one type of food. Thus, the main bulk of the muriqui diet comes from fruit, but muriquis also eat leaves, as you see here. (f) Langurs are folivores. Here, Hanuman langurs at Ramnagur forage for water plants. (Photographs courtesy of: a, Lynne Isbell; b, Susan Perry; c, John Mitani; e, Carlão Limeira; f, Carola Borries.)

frugivores or insectivores. Thus, it is not surprising to find that folivores generally have smaller home ranges than frugivores or insectivores.

Activity Patterns

The fact that many prosimian species are nocturnal suggests that primates evolved from a nocturnal ancestor.

Most primate species are active either during the day (diurnal) or at night (nocturnal), although a few species are active at intervals throughout the day and night **(cathemeral).** Since all monkeys and apes (except the owl monkey) are diurnal, and a substantial number of prosimians are nocturnal, it seems likely that primates originally

ACTIVITY PATTERNS

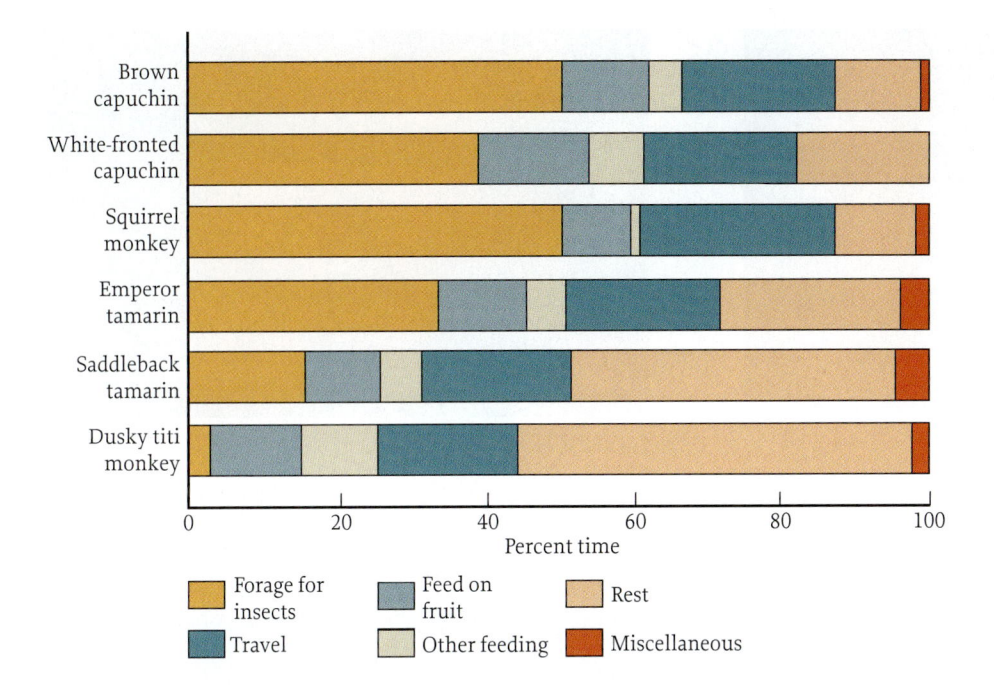

FIGURE 6.9

The amount of time that animals devote to various types of activities is called a time budget. Time budgets of different species vary considerably. These six monkey species all live in a tropical rain forest in Manu National Park in Peru.

evolved from a nocturnal ancestor. Nocturnal primate species tend to be smaller, more solitary, and more exclusively arboreal than diurnal species. They also rely more heavily on olfactory signals than do diurnal species.

Primate activity patterns show regularity in seasonal and daily cycles.

Primates spend the majority of their time feeding, moving around their home ranges, and resting (Figures 6.9 and 6.10). Relatively small portions of each day are spent grooming, playing, fighting, or mating (Figure 6.11). The proportion of time devoted to various activities is influenced to some extent by ecological conditions. For primates living in seasonal habitats, for example, the dry season is often a time of scarce resources, and it is harder to find adequate amounts of appropriate types of food. In some cases, this means that the proportion of time spent feeding and traveling increases during the dry season, while the proportion of time spent resting and socializing decreases.

Primate activity also shows regular patterns over the course of the day. When they wake up, their stomachs are empty, so the first order of the day is to visit a feeding site. Much of the morning is spent eating and moving between feeding sites. As the sun moves directly overhead and the temperature rises, most species settle down in a shady spot to rest, socialize, and digest their morning meals (Figure 6.12). Later in the afternoon, they resume feeding. Before dusk, they move to the night's sleeping site; some species habitually sleep in the same trees every night, while others have multiple sleeping sites within their ranges (Figure 6.13).

FIGURE 6.10

All diurnal primates, like this capuchin monkey, spend some part of each day resting. (Photograph courtesy of Susan Perry.)

FIGURE 6.11
Immature monkeys spend much of their "free" time playing. These patas monkeys are play wrestling. (Photograph courtesy of Lynne Isbell.)

FIGURE 6.12
Gorillas often rest in close proximity to other group members and socialize during a midday rest period.

FIGURE 6.13
Like most other primates, baboons sleep perched on the branches of trees at night. These baboons are descending from their sleeping trees shortly after dawn.

Ranging Behavior

All primates have home ranges, but only some species are territorial—defending their home range against incursions by other members of their species.

In virtually all primate species, groups range over a relatively fixed area, and members of a given group can be consistently found in a particular area over time. These areas are called home ranges, and they contain all of the resources that group members exploit in feeding, resting, and sleeping. However, the extent of overlap among adjacent home ranges and the nature of interactions with members of neighboring groups or strangers vary considerably among species. Some primate species, like gibbons, maintain exclusive access to fixed areas, called **territories** (Figure 6.14a). Territory residents regularly advertise their presence by vocalizing, and they aggressively protect the boundaries of their territories from encroachment by outsiders (Figure 6.15). While some territorial birds defend only their nest sites, primate territories contain all of the sites at which the residents feed, rest, and sleep and the areas in which they travel. Thus, among territorial primates, the boundaries for the territory are essentially the same as for their home range, and territories do not overlap.

Nonterritorial species, like squirrel monkeys and long-tailed macaques, establish home ranges that overlap considerably with those of neighboring groups (Figure 6.14b). When members of neighboring nonterritorial groups meet, they may fight (Figure 6.16), exclude members of lower-ranking groups from resources, avoid one another, or mingle peacefully together. This last option is unusual, but in some species, adult females sexually solicit males from other groups, males attempt to mate with females from other groups, and juveniles from neighboring groups play together when their groups are in proximity.

The two main functions suggested for territoriality are resource defense and mate defense.

In order to understand why some primate species defend their home ranges from intruders and others do not, we need to think about the possible costs and benefits associated with defending resources from conspecifics. Costs and benefits are measured in terms of the impact on the individual's ability to survive and reproduce successfully. Territoriality is beneficial because it prevents outsiders from exploiting the limited resources within a territory. At the same time, however, territoriality is costly because the residents must be constantly vigilant against intruders, regularly adver-

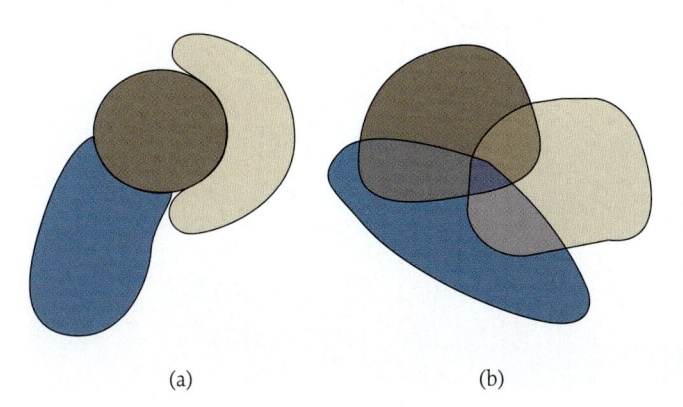

FIGURE 6.14

(a) Some primates defend the boundaries of territories (a schematic map of three territories is shown here) and do not tolerate intrusions from neighbors or strangers of the same species. (b) Other primates occupy fixed home ranges (three shown here) but do not defend their borders against members of the same species. When two groups meet in areas of home range overlap, one may defer to the other, they may fight, or they may mingle peacefully together.

(a)   (b)

FIGURE 6.15
Gibbons perform complex vocal duets as part of territorial defense.

tise their presence, and be prepared to defend their ranges against encroachment. Territoriality is expected to occur only when the benefits of maintaining exclusive access to a particular piece of land outweigh the costs of protecting these benefits.

When will the benefits of territoriality exceed the costs? The answer to this question depends in part on the kind of resources that individuals need in order to survive and reproduce successfully, and in part on the way these resources are distributed spatially and seasonally. For reasons we will discuss more fully in Chapter 7, the reproductive strategies of mammalian males and females generally differ. In most cases, female reproductive success depends mainly on getting enough to eat for themselves and their dependent offspring, while males' reproductive success depends mainly on their ability to mate with females. As a consequence, females are more concerned about access to food, while males are more interested in access to females. Thus, territoriality has two different functions. Sometimes females defend food resources, or males defend food resources on their behalf. Other times, males defend groups of females against incursions by other males. In primates, both resource defense and mate defense seem to have influenced the evolution of territoriality.

Resource-defense territoriality occurs when resources are not only limited but also clumped and defendable.

Territoriality occurs when resources are economically defensible, meaning that resources are limited in abundance but occur within an area that can be defended with a reasonable amount of effort. When food resources are distributed over a wide area, it is costly to detect and evict intruders, so territoriality does not pay. Similarly, when resources are readily available, it makes little sense to defend them (Figure 6.17a). If, on the other hand, valuable resources are clumped within a small area, territoriality can be favored by natural selection (Figure 6.17b).

FIGURE 6.16
Some primates defend fixed territories. Others react aggressively when they meet members of other groups. Here, a group of Hanuman langurs attempts to repel an intruder. (Photograph courtesy of Carola Borries.)

PREDATION

FIGURE 6.17
(a) When resources are evenly distributed, like these *Ramphicarpa montana* flowers, there is little point in defending access to them. (b) When resources are clumped together, like these acacia flowers, then animals may profit from driving away competitors.

Mate defense also plays a role in the evolution of territoriality in some primate species.

Although territoriality seems to be linked to the defendability of resources in many species, there are also species in which resource defense does not seem to be the primary factor favoring territoriality. This conclusion is based on the observation that in a number of species, males are active in intergroup encounters while females seem largely indifferent to the presence of intruders. This pattern is characteristic of gorillas, red colobus monkeys, many Southeast Asian langurs, and some species of lesser apes. Since males' reproductive success is influenced by their access to females, when males are the principal actors in territorial encounters, it seems likely that the primary function of territoriality is to defend access to mates, not food resources. By the same token, since females' ability to reproduce is related to their access to food resources, it seems likely that if females don't participate in territorial disputes, then resource defense is not the central factor determining the nature of intergroup encounters.

Predation

Predation is believed to be a significant source of mortality among primates, but direct evidence of predation is normally difficult to obtain.

Primates are hunted by a wide range of predators, including pythons, raptors, crocodiles, leopards, lions, tigers, and humans (Figure 6.18). In Madagascar, large lemurs are preyed upon by fossas, puma-like carnivores. Primates are also preyed on by other primates. Chimpanzees, for example, hunt red colobus monkeys, and baboons prey on vervet monkeys.

In primate populations, the estimated rates of predation vary from less than 1% of the population per year to more than 15% of the population per year. The available data suggest that small-bodied primates are more vulnerable to predation than larger ones are (Figure 6.19), and immature primates are generally more susceptible to predation than adults are. These data are not very solid because systematic information about predation is quite hard to come by. Predation is very hard to observe directly since most predators avoid close contact with humans, and the presence of human observers probably deters attacks. Some predators, like leopards, generally hunt at night when most researchers are asleep. In most cases, predation is inferred when a

160   PRIMATE ECOLOGY

(a)   (b)   (c)

(e)

FIGURE 6.18

Primates are preyed upon by a variety of predators, including (a) pythons, (b) lions, (c) leopards, (d) martial eagles, and (e) crocodiles.

(d)

healthy animal who is unlikely to have left the group abruptly vanishes without a trace (Figure 6.20). Such inferences are, of course, subject to error.

Another approach is to study the predators themselves. John Mitani of the University of Michigan and his colleagues studied the predatory behavior of crowned hawk eagles (Figure 6.21) in Kibale Forest in Uganda by collecting the prey remains under the eagles' nests. Crowned hawk eagles are formidable predators; although they weigh only 3 to 4 kg (6.6 to 8.8 lb), they can take prey that weigh up to 30 kg (66 lb). In Kibale, crowned hawk eagles prey on all of the primates in the forest except chimpanzees, which weigh 30 to 50 kg (66 to 110 lb). Monkeys make up about 80% of the eagles' diet. Mitani and his colleagues estimated that approximately 2% of the total monkey population in Kibale is killed by hawk eagles each year.

FIGURE 6.19

Small primates are more vulnerable to predators, but even large primates are vigilant. (a) This squirrel monkey is scanning the canopy for potential predators. (b) A male rhesus monkey scans the ground below for possible dangers. (a, photograph courtesy of Sue Boinski.)

(a)   (b)

(a)   (b)   (c)

FIGURE 6.20

In some cases, researchers are able to confirm predation. Here, an adult female baboon in the Okavango Delta, Botswana, was killed by a leopard. You can see (a) the depression in the sand that was made when the leopard dragged the female's body out of the sleeping tree and across a small sandy clearing, (b) the leopard's footprints beside the drag marks, and (c) the remains of the female the following morning—her jaw, bits of her skull, and clumps of hair.

Primates have evolved an array of defenses against predators.

Many primates give alarm calls when they sight potential predators, and some species have specific vocalizations for particular predators. Vervet monkeys, for example, give different calls when they are alerted to the presence of leopards, small carnivores, eagles, snakes, baboons, and unfamiliar humans. In many species, the most common response to predators is to flee or take cover. Small primates sometimes try to conceal themselves from predators, while larger ones may confront potential predators. For example, slow-moving pottos encountering snakes fall to the ground, move a short distance, and freeze. At some sites, adult red colobus monkeys aggressively attack chimpanzees who hunt their infants.

Another antipredator strategy that primates adopt is to associate with members of other primate species. In the Taï Forest of Ivory Coast, a number of monkey species share the canopy and form regular associations with one another. For example, groups of red colobus monkeys spend approximately half their time with groups of Diana monkeys. Ronald Noë and Rdeouan Bshary of the Université Louis Pasteur in Strasbourg, France, have shown that both species benefit when they are together. Red colobus monkeys sometimes detect eagles before Diana monkeys do, and Diana monkeys often detect terrestrial predators before the red colobus do. As a result, red colobus monkeys are less vigilant about dangers from terrestial predators when they are with Diana monkeys, and Diana monkeys are less vigilant about aerial predators when they are with the red colobus.

FIGURE 6.21

Crowned hawk eagles specialize in hunting monkeys. Primates make up 80% of their diet in the Kibale Forest of Uganda. There, crowned hawk eagles prey on all of the endemic primate species except chimpanzees.

Primate Sociality

Most primates, except for orangutans and some prosimians, spend most of their lives in stable groups of familiar individuals.

Although nearly all primates live in groups (Figure 6.22), not all groups are alike. The social organization among monkeys and apes encompasses great diversity, ranging from monogamous, pair-bonded groups of gibbons and owl monkeys, to polyan-

FIGURE 6.22

Most primates live in social groups. (a) Baboons often live in groups of 30 to 60 individuals. All members of the group recognize each other as individuals. (b) Ring-tailed lemurs live in female-bonded groups, from which males disperse at puberty.

drous groups of tamarins; one-male groups of howlers and blue monkeys; multimale, multifemale groups of cebus monkeys and macaques; and structured communities of gelada and hamadryas baboons (Box 6.2). The great diversity in primate social organization has prompted researchers to ask a number of related questions: Why do primates live in groups? How should groups be structured? How big should groups be?

Sociality is not unique to primates—killer whales, wolves, zebra, mongoose, and elephants (Figure 6.24) are some of the many animals that live in cohesive social groups. However, group life is not common among mammals. In many mammalian species, adult males and females occupy separate home ranges or territories and meet only for brief periods to court and mate. Females raise their young without further participation from males. Thus, we need to consider why sociality has evolved among primates.

Social life has both costs and benefits.

Sociality reflects a dynamic balance between the advantages and disadvantages of living in close proximity to conspecifics. Sociality is beneficial because primates who live in social groups are better able to acquire and control resources and are less vulnerable to predators. Animals who live in groups can chase lone individuals away from feeding trees and can protect their own access to food and other resources against smaller numbers of intruders. Grouping also provides safety from predators because of the three *D*'s—detection, deterrence, and dilution. Animals in groups are more likely to detect predators because there are more pairs of eyes on the lookout for predators. Animals in groups are also more effective in deterring predators by actively mobbing or chasing them away. Finally, the threat of predation to any single individual is diluted when predators strike at random. If there are two animals in a group, and a predator strikes, each has a 50% chance of being eaten. If there are 10 individuals, the individual risk is decreased to 10%.

Although there are important benefits associated with sociality, there are equally important costs. Group-living animals may encounter more competition over access

Box 6.2
Forms of Social Groups among Primates

It should be obvious by now that while most primates are social, their groups vary in size and composition. One way to order this diversity is to classify primate social systems according to the mating and residence patterns of females and males. The major forms of primate social systems are described here and diagrammed in Figure 6.23:

Solitary Females maintain separate home ranges or territories and associate mainly with their dependent offspring. Males may establish their own territories or home ranges, or they may defend the ranges of several adult females from incursions by other males. With the exception of orangutans, all of the solitary primates are prosimians.

Monogamy One male and one female form a pair-bond and share a territory with their immature offspring. Monogamy is characteristic of gibbons, some of the small New World monkeys, and a few prosimian species.

Polyandry One female is paired with two or more males. They share a territory or home range with their offspring. Polyandry may occur among some of the marmosets and tamarins.

Polygyny, one-male Groups are composed of several adult females, a single resident male, and immature individuals. (We refer to them in the text as one-male, multifemale groups.) In species with this form of polygyny, males that do not reside in bisexual groups often form all-male groups. These all-male groups may mount vigorous assaults on resident males of bisexual groups in an attempt to oust residents. Dispersal patterns are variable in species that form one-male, multifemale groups. One-male, multifemale groups are common among howlers, langurs, and gelada baboons.

Polygyny, multimale, multifemale Groups are composed of several adult males, adult females, and immature animals.

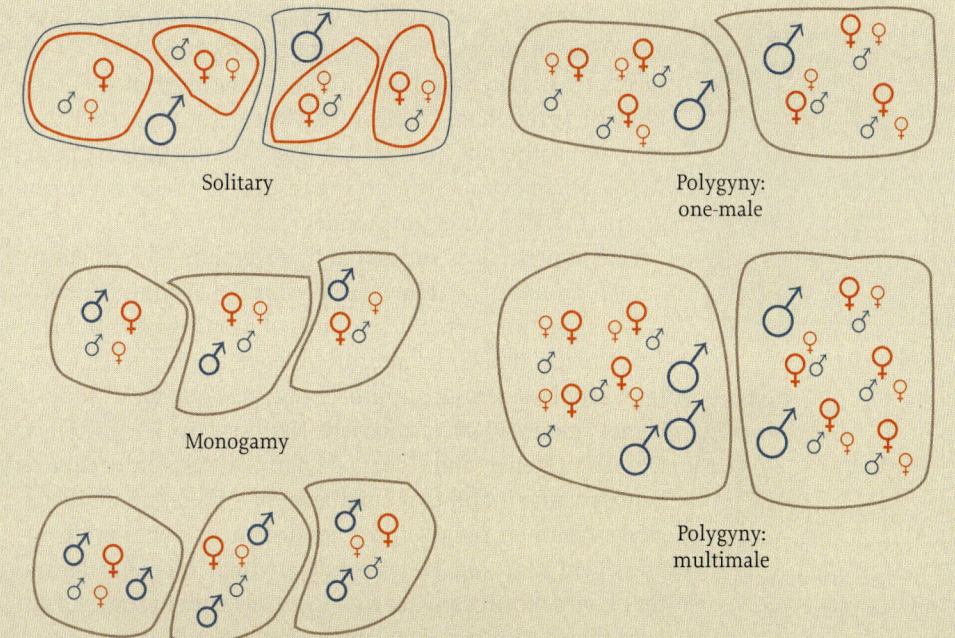

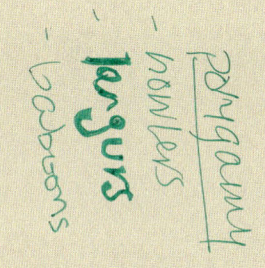

FIGURE 6.23

The major types of social groups that primates form are diagrammed here. When males and females share their home ranges, their home ranges are drawn in lavender. When the ranges of the two sexes differ, male home ranges are drawn in blue and female home ranges are drawn in magenta. The size of the male and female symbols reflect the degree of sexual dimorphism among males and females.

(We refer to them in the text as multimale, multifemale groups.) Most such groups are female-bonded. Dispersing males may move directly between bisexual groups, remain solitary for short periods of time, or join all-male groups. Multimale, multifemale groups are characteristic of macaques, baboons, vervets, squirrel monkeys, capuchin monkeys, and some colobines.

This taxonomy of social groups provides one way to classify primate social organization, but readers should recognize that these categories are idealized descriptions of residence and mating patterns. The reality is inevitably more complicated. For example, in some monogamous groups, partners sometimes copulate with nonresidents or neighbors, and in species that normally form multimale groups, there may be some groups that have only one adult male. Moreover, some species do not fit neatly within this classification system. Female chimpanzees and spider monkeys occupy individual home ranges and spend much of their time alone or with their dependent offspring. In these species, a number of males jointly defend the home ranges of a number of females. We call these fission-fusion groups because party size is variable. Hamadryas and gelada baboons form one-male, multifemale units, but several one-male, multifemale units collectively belong to larger aggregations.

to food and mates, become more vulnerable to disease, and face various hazards from conspecifics (such as cannibalism, cuckoldry, incest, or infanticide).

The size and composition of the groups that we see in nature are expected to reflect a compromise between the costs and benefits of sociality for individuals. The magnitude of these costs and benefits is influenced by both social and ecological factors. Thus, the study of how natural selection shapes social organization in response to ecological pressures is called **socioecology**.

Primatologists are divided over whether feeding competition or predation is the primary factor favoring sociality among primates.

Joint defense of food resources will be profitable when (1) food items are relatively valuable, (2) food resources are clumped in time and space, and (3) there is enough food within the patch to meet the needs of several individuals. Many primates feed preferentially on fruit, a food that meets these criteria (Figure 6.25). Taking note of this fact, Richard Wrangham of Harvard University suggested that primates gather together into groups because they are more successful in defending access to resources when they are in groups than when they are alone.

This hypothesis is supported by several lines of evidence. Larger groups generally prevail during encounters with smaller groups. Even when groups of monkeys do not meet face to face, smaller groups often leave feeding trees when they hear larger groups approaching. In a long-term study of vervet monkeys in Amboseli National Park in Kenya, Dorothy Cheney and Robert Seyfarth of the University of Pennsylvania found that small groups suffered more incursions into their ranges and defended their ranges more aggressively than did larger groups, which probably means that it was more costly for members of smaller groups to defend their territories. The larger groups also had bigger and more desirable territories than smaller groups did. These dif-

FIGURE 6.24
Sociality is not restricted to primates. For example, the social organization of elephants is highly structured. The basic social unit is the family group, which consists of one or more related adult females and their offspring. Certain family groups are frequently found together and form bond groups. Members of certain bond groups tend to associate with one another and are clustered into clans.

ferences in territorial defense and territory quality were reflected in females' life histories, as females in the small groups had lower survivorship than did females in the large groups.

Although Wrangham's idea is cogent and is supported by some evidence, Duke University primatologist Carel van Schaik challenged the belief that sociality is favored because it enhances resource defense. He pointed out that there is considerable competition over access to food *within* primate groups, and this may outweigh any advantages gained in competition *between* groups. For example, Charles Janson of the State University of New York at Stony Brook has assessed the extent of competition within and between groups of brown capuchins in Peru's Manu National Park (Figure 6.26). In this population, monkeys compete actively and often with other group members over access to food; the rate of aggression increases as group size increases (Figure 6.27). In contrast, intergroup competition was limited to relatively brief and infrequent contests over access to fruiting fig trees. Energy intake varied by 37% among individuals within the group, whereas energy intake varied by only 3% between groups that were most and least successful in intergroup competition.

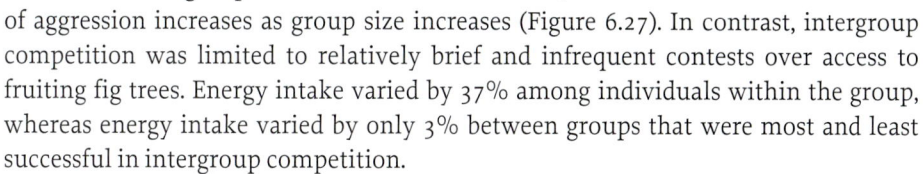

The brown capuchin data suggest that if primates wanted to avoid competition with conspecifics, they might be better off avoiding them altogether. However, this option is dangerous for primates because predators present a serious threat to lone individuals. Thus, van Schaik suggested that primates live in groups because groups provide greater safety from predators. The predation model is difficult to test because predation is so hard to study as we explained earlier. However, several lines of evidence support the idea that sociality reduces predation risk. First, macaques living on islands without predators form smaller groups than do their more vulnerable conspecifics on the mainland. Second, most of the species that typically do not forage in groups (for example, spider monkeys, chimpanzees, and orangutans) are large animals and apparently face little danger from predators (Figure 6.28). Third, some of the species that regularly form mixed-species groups in the Taï Forest do not associate with each other at sites where predators are absent.

FIGURE 6.25

Femal baboons feed together on *Salvadora persica* fruit in Amboseli, Kenya.

FIGURE 6.26

A brown capuchin monkey feeds in Peru's Manu National Park. (Photograph courtesy of Charles Janson.)

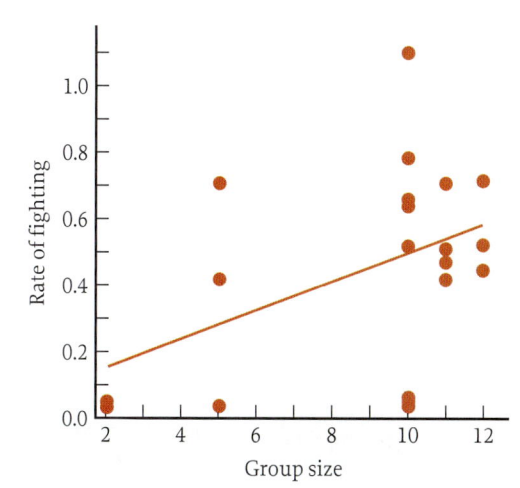

FIGURE 6.27

One possible cost of living in social groups is increased competition among group members over access to resources. For example, the rate of aggression increases as group size increases among capuchin monkeys in Peru's Manu National Park.

FIGURE 6.28
Orangutans do not live in social groups. They are the only solitary anthropoid species.

Finally, in a number of species, vigilance rates decline when group members are nearby. For example, Adrian Treves and his colleagues from the University of Wisconsin found that black howler monkeys spend less time scanning for predators when companions are nearby (Figure 6.29).

Although these data suggest that grouping reduces vulnerability to predation, some data don't fit this hypothesis. Thus, in a recent review of the literature on vigilance in primates, Treves showed that vigilance does not decline smoothly as group size increases, perhaps because predation is not the only hazard that group-living primates face.

The jury is still out on whether resource competition or predation was the primary factor favoring the evolution of sociality in primates. However, many primatologists are convinced that the nature of resource competition affects the behavioral strategies of primates, particularly females, and influences the composition of primate groups. Females come first in this scenario because females' fitness depends mainly on their nutritional status. Thus, well-nourished females grow faster, mature earlier, and have higher fertility rates than do poorly nourished females. In contrast, males' fitness depends primarily on their ability to obtain access to fertile females, not on their nutritional status. (We will discuss male and female reproductive strategies more fully in Chapter 7.)

THE DISTRIBUTION OF FEMALES

Ecologists distinguish between two kinds of resource competition, scramble competition and contest competition.

Scramble competition occurs when resources are distributed evenly across the landscape. Animals cannot effectively monopolize access to resources when they are distributed this way, so they do not compete over them directly. (Think about what happens when a piñata is broken open and all the candy rains down—everyone scrambles for the candy, but they don't fight over any particular item.) **Contest competition** occurs when resources are limited and can be monopolized profitably, generating direct confrontations over access to them. (Here, think about what happens in musical chairs—when the music stops, everyone scrambles for a chair, knowing that there are not enough to go around.) Both scramble and contest competition can occur within groups and between groups.

FIGURE 6.29
Primates in groups may be less vulnerable to predators and thus may be able to reduce their levels of vigilance. Black howler monkeys are less vigilant when other monkeys are in the same tree than when they are alone.

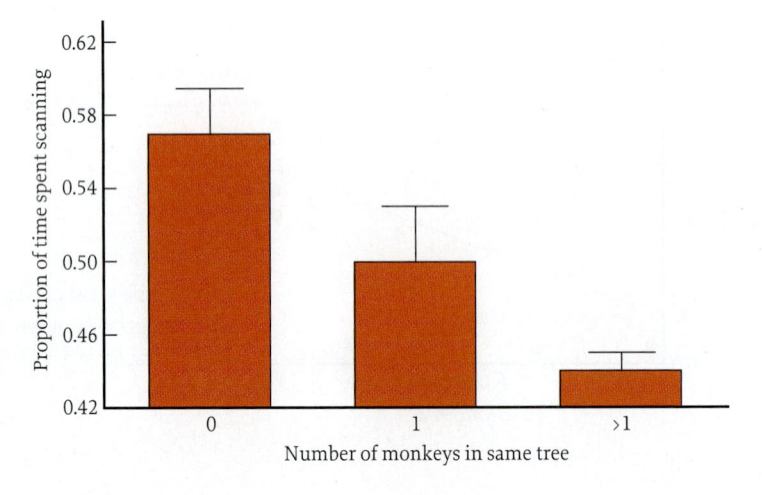

PRIMATE SOCIALITY

Resource competition is expected to generate dominance relationships.

In many animal species, ranging from crickets to chickens to chimpanzees, competitive encounters within pairs of individuals are common. The outcome of these contests may be related to the participants' relative size, strength, experience, or willingness to fight. In many species, for example, larger and heavier individuals regularly defeat smaller individuals. If there are real differences in power (based on size, weight, experience, or aggressiveness) between individuals, then we would expect the outcomes of dominance contests to be more or less the same from day to day. This is often the case. When dominance interactions between two individuals have predictable outcomes, we say that a **dominance** relationship has been established. We can plot the outcome of pairwise dominance contests in a matrix and construct **dominance hierarchies** to generate dominance rankings for individuals (Box 6.3).

When contest competition within groups is stronger than contest competition between groups, females will remain in their natal groups and cooperate with their relatives in contests with unrelated females in their group over resources. This leads to dominance hierarchies in which related females have adjacent ranks.

Socioecological models consider the effects of within- and between-group competition on primate females. The model that we outline here was developed by primatologist Elisabeth Sterck of Utrecht University, David Watts of Yale University, and Carel van Schaik. Don't be put off by its complexity; many of the ideas that we introduce here will come up again in Chapters 7 and 8.

Sterck and her colleagues predict that when within-group contest competition is the primary form of competition, females' ability to control access to resources will be a function of their dominance rank. All other things being equal, high-ranking females are routinely able to exclude lower-ranking females from resources. This simple fact is expected to have a number of important consequences for primate females.

If dominance rank influences females' access to resources, and this in turn affects their fitness, females should strive for high rank. Moreover, females may benefit from helping their relatives, who share some of their genes, gain high rank. (We will explain how kinship facilitates cooperation more fully in Chapter 8.) Because there is power in numbers, females can help their relatives win dominance contests by coming to their aid when they are challenged and supporting them when they initiate attacks. If females help their relatives in dominance encounters, then females will come to be able to defeat everyone that their relatives can defeat. We will see in Chapter 8 that this leads to the formation of linear dominance hierarchies in which maternal kin occupy adjacent ranks.

If kin play an important role in the acquisition and maintenance of rank, then there are strong incentives for females to remain in their natal (birth) groups with their mothers, sisters, aunts, and cousins. We call this female **philopatry.** Finally, friendly interactions, like grooming and sitting close together, may enhance the quality of social relationships and reinforce alliances. If so, females are expected to interact preferentially with their coalition partners.

When between-group contest competition is stronger than within-group contest competition, dominance will be less important, but female philopatry will still be favored.

In contrast, when between-group contest competition is strong and within-group contest competition is negligible, females' ability to gain access to resources will depend on the size of the social group that they live in, not their own dominance rank.

Box 6.3
Dominance Hierarchies

When dominance interactions have predictable outcomes, we can assign dominance rankings to individuals. Consider the four hypothetical females which we shall call Blue, Turquoise, Green, and Lavender in Figure 6.30a. Blue always beats Turquoise, Green, and Lavender. Turquoise never beats Blue but always beats Green and Lavender. Green never beats Blue or Turquoise but always beats Lavender. Poor Lavender never beats anybody. We can summarize the outcome of these confrontations between pairs of females in a **dominance matrix** like the one in Figure 6.30b, and we can use the data to assign numerical ranks to the females. In this case, Blue ranks first, Turquoise second, Green third, and Lavender fourth. When females can defeat all the females ranked below them and none of the females ranked above them, dominance relationships are said to be **transitive**. When the relationships within all sets of three individuals (trios) are transitive, the hierarchy is linear.

FIGURE 6.30
(a) Suppose that four hypothetical females, named Blue, Turquoise, Green, and Lavender, have the following transitive dominance relationships: Blue defeats the other three in dominance contests. Turquoise cannot defeat Blue, but can defeat Green and Lavender. Green loses to Blue and Turquoise, but is able to defeat Lavender. Lavender can't defeat anyone. (b) The results of data like those in part a are often tabulated in a dominance matrix, with the winners listed down the left side and the losers across the top. The value in each cell of the matrix represents the number of times one female defeated the other. Here, blue defeated Turquoise 10 times, and Green defeated Lavender 11 times. There are no entries below the diagonal because females were never defeated by lower-ranking females.

This will change the dynamics of social interactions within groups in a number of ways. First, it will weaken competition for access to resources within groups, so females will not strive for high rank and dominance relationships will tend to be more egalitarian. Second, there will be less incentive to support kin when they are involved in disputes with other group members, but females will rely on coalitionary support in intergroup encounters from all group members. Because females benefit more from supporting kin than nonkin, female philopatry will again be favored. However, because females' fitness depends on collective group action, females are expected to cultivate more uniform social relationships with other group members.

PRIMATE SOCIALITY

When both within-group and between-group contest competitions are strong, the result will be intermediate between the previous scenarios.

The third possibility is that there will be strong contest competition within and between groups. In this case, females' fitness will depend on their ability to win competitive encounters with other group members, as well as their ability to compete as a group with members of neighboring groups. This will produce groups that combine the highly nepotistic structure that we expect to find when within-group contest competition predominates, with the more egalitarian structure that we expect to find when between-group contest competition prevails.

Finally, regimes in which only scramble competition prevails both within and between groups are possible. In these cases, there will be little reason for females to establish dominance relationships, to form alliances with other females, or to participate in collective activities. Thus, ecological pressures will not guide females' behavior in any particular direction. Most researchers assume that this is a very atypical situation for primate females to find themselves in.

Empirical work supports the predictions of socioecological models.

In the 1990s, many researchers set to work testing the predictions derived from these socioecological models. This body of work confirmed many of the predictions. For example:

- Sue Boinski of the University of Florida and her colleagues compared two closely related species of squirrel monkeys in Costa Rica and Peru that had similar diets, formed groups of similar sizes, and experienced similarly high levels of predation. The two species differed mainly in the form of within-group competition over access to food resources. As predicted, the species that experienced high within-group contest competition formed stable, linear dominance hierarchies, developed kin-based coalitions, and exhibited strict female philopatry. In the species that experienced low levels of within-group contest competition, a dominance hierarchy was not detected among females, female coalitions were not observed, and females sometimes left their natal groups.
- Lynn Isbell and her colleagues at the University of California, Davis, have compared the feeding ecology and social behavior of female monkeys in two closely related species, patas and vervets (Figure 6.31). Patas rely more heavily than vervets on arthropods and other food items that require little handling time and are consumed very quickly. This is a situation in which within-group scramble competition should prevail. On the other hand, vervets rely more heavily than patas on foods such as gums, seeds, and fruits that have longer processing times and take

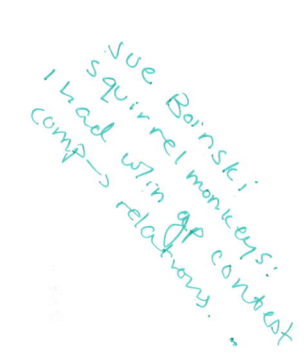

FIGURE 6.31

Vervets and patas are closely related but differ in key features of their feeding ecology. Socioecological models predict that female patas monkeys (pictured on the left) will have more egalitarian dominance hierarchies and less well-defined social bonds than will female vervets (pictured on the right).

FIGURE 6.32

Hanuman langurs at Ramnagar encounter strong within-group contest competition and form stable, linear dominance hierarchies. (Photograph courtesy of Carola Borries.)

longer to eat. Within-group contest competition would be expected to be more important for vervets. Patas have less linear dominance hierarchies and weaker social bonds than vervets do, as we would expect.

- Andreas Koenig, now at the State University of New York at Stony Brook, and his colleagues compared two populations of Hanuman langurs living at different sites (Figure 6.32). The langurs of Ramnagar in southern Nepal relied heavily on just three food plants. These plants were low in abundance and clumped in their distribution, conditions that are expected to lead to within-group contest competition. The langurs of Kanha in India relied on a larger array of plant foods, and one of their most highly preferred foods is superabundant, conditions that are expected to lead to within- and between-group scramble competition. The langurs of Ramnagar behave much as expected when within-group contest competition is strong: they have linear dominance hierarchies and females are philopatric. The langurs of Kanha have poorly developed dominance hierarchies, and females sometimes emigrate from their natal groups.

Not all of the observed variations in social organization and behavior fit predictions derived from socioecological models, perhaps because phylogeny constrains social evolution.

Although many of the comparisons of two closely related species provide support for socioecological models, we can also find examples that don't fit the model well. For instance, baboons occupy an extremely diverse range of habitats across Africa, but the basic structure of their social organization remains remarkably constant across habitats and groups.

The role of the environment in shaping social organization has also been challenged by comparative studies that take phylogeny explicitly into account. Anthony di Fiore of New York University and Drew Rendall of Lethbridge University point out that most of the characteristics associated with within-group contest competition, including linear dominance hierarchies, nepotistic alliances, and female philopatry, are found in almost all of the extant cercopithecine species, even though they now occupy an extremely diverse range of habitats. These findings suggest that there is considerable inertia in social evolution. The social systems categorized by socioecologists may represent different peaks in the adaptive landscape; once a species evolves one kind of social organization, it may be extremely difficult to shift to another kind, even if ecological conditions change. If this is true, then it seriously constrains the possible paths for change in social organization.

THE DISTRIBUTION OF MALES

In socioecological models, ecological factors shape the distribution of females, and males go where females are.

As we noted earlier, male reproductive fitness depends mainly on access to females, not food. So males go where females are. When females live in groups, they become a

PRIMATE SOCIALITY

defensible resource for males. The more females there are, the more difficult it is for a single male to monopolize them. Moreover, if females are sexually receptive at the same time, it is more difficult for a single male to monopolize them. Comparative analyses indicate that the number of males in primate groups is generally correlated with the number of females present and with the extent of synchrony in female receptivity (Figure 6.33).

The distribution of females explains some but not all of the variation in the number of males in social groups. Some researchers think that the distribution of males in social groups is also influenced by ecological factors, particularly pressure from predators. Robert Hill and Phyllis Lee of Cambridge University conducted a systematic survey of Old World monkeys in an effort to assess the effects of predation risk on the distribution of males. They found that groups that faced the highest risks of attacks by predators lived in the largest groups and had the most males. Thus, the distribution of males may be a joint product of the distribution of females and the intensity of predation pressures.

The nature of resource competition also influences male dispersal strategies because inbreeding is disadvantageous.

Notice that female philopatry is expected to occur in three of the four competitive regimes described earlier. This has important consequences for male life histories. The fitness of males depends largely on the availability of unrelated females. The presence of unrelated females is important because in most species, mating with close kin (**inbreeding**) reduces the viability of offspring by increasing the probability that deleterious recessive traits will be expressed. Surveys of captive primate populations have shown that offspring of closely related parents are less likely to survive than are offspring of unrelated parents, and close kin generally avoid mating with one another. In all primate species studied to date, members of one or both sexes typically leave their natal groups near the time of sexual maturity. When females are philopatric, males must disperse. As you will see in Chapter 7, dispersal imposes significant costs on males.

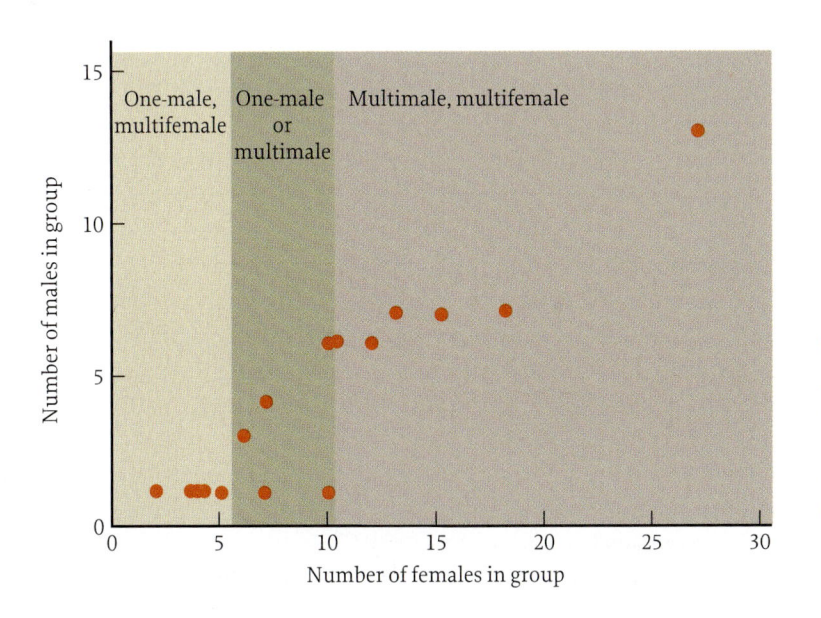

FIGURE 6.33

Among cercopithecine primate species, the number of females in social groups is positively correlated with the number of males in the same groups. Groups with only a few adult females generally contain only one adult male, while groups with many females contain several males adult males.

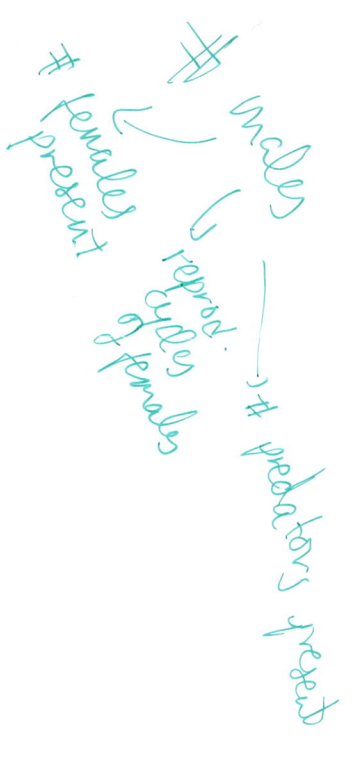

Further Reading

Boinski, S., and P. A. Garber. 2000. *On the Move: How and Why Animals Travel in Groups.* University of Chicago Press, Chicago.

Janson, C. H. 2000. Primate socioecology: the end of a golden era. *Evolutionary Anthropology* 9:73–86.

Oates, J. F. 1986. Food distribution and foraging behavior. Pp. 197–209 in *Primate Societies*, ed. by B. B. Smuts, D. L. Cheney, R. M. Seyfarth, R. W. Wrangham, and T. T. Struhsaker. University of Chicago Press, Chicago.

Richard, A. 1985. *Primates in Nature*, W. H. Freeman, New York, chaps. 4 and 5.

Sterck, E. H. M., D. P. Watts, and C. P. van Schaik. 1997. The evolution of social relationships in nonhuman primates. *Behavioral Ecology and Sociobiology* 41:291–309.

van Schaik, C. P. 1989. The ecology of social relationships amongst female primates. Pp. 195–218 in *Comparative Socioecology: The Behavioural Ecology of Humans and Other Mammals*, ed. by V. Standon and R. A. Foley. Oxford University Press, Oxford.

Wrangham, R. W. 1980. An ecological model of female-bonded primate groups. *Behaviour* 75:262–300.

Study Questions

1. Large primates often subsist on low-quality food such as leaves, while small primates specialize in high-quality foods such as fruit and insects. Why is body size associated with dietary quality in this way?
2. For folivores, tropical forests seem to provide an abundant and constant supply of food. Why is this not an accurate assessment?
3. Territorial primates do not have to share access to food, sleeping sites, mates, and other resources with members of other groups. Given that territoriality reduces the extent of competition over resources, why aren't all primates territorial?
4. Territoriality is often linked to group size, day range, and diet. What is the nature of the association, and why does the association occur?
5. Most primates specialize in one type of food, such as fruit, leaves, or insects. What benefits might such specializations have? What costs might be associated with specialization?
6. Nocturnal primates are smaller, more solitary, and more arboreal than diurnal primates. What might the reason(s) be for this pattern?
7. Some primatologists have predicted that there should be a linear effect of group size on vigilance rates, if sociality evolved as a defense against predation. This means that every increase in the number of animals in the group should be associated with a decrease in the proportion of time that animals spend in vigilance activities. Many studies have failed to show a linear effect. How would you interpret these results? What do the results tell you about the selective forces that shape sociality in primates?
8. Sociality is a relatively uncommon feature in nature. What are the potential advantages and disadvantages of living in social groups? Why are (virtually all) primates social?
9. A number of studies have attempted to test socioecological models by comparing two closely related species. What is the logic underlying such studies? What would be another approach?
10. Males are largely left out of socioecological models. Why is this the case? What influences the distribution of males?

CHAPTER 7

Primate Mating Systems

THE LANGUAGE OF ADAPTIVE EXPLANATIONS
THE EVOLUTION OF REPRODUCTIVE STRATEGIES
REPRODUCTIVE STRATEGIES OF FEMALES
    SOURCES OF VARIATION IN FEMALE REPRODUCTIVE PERFORMANCE
    REPRODUCTIVE TRADEOFFS
SEXUAL SELECTION AND MALE MATING STRATEGIES
    INTRASEXUAL SELECTION
    INTERSEXUAL SELECTION
MALE REPRODUCTIVE TACTICS
    INVESTING MALES
    MALE-MALE COMPETITION IN NONMONOGAMOUS GROUPS
    INFANTICIDE
    PATERNAL CARE IN NONMONOGAMOUS GROUPS
    FEMALE MATE CHOICE

Reproduction is the central act in the life of every living thing. Primates perform a dizzying variety of behaviors—gibbons fill the forest with their haunting duets, baboons threaten and posture in their struggle for dominance over other members of their group, and chimpanzees use carefully selected stone hammers to crack open tough nuts. But all of these behaviors evolved for a single ultimate purpose—reproduction. According to Darwin's theory, complex adaptations exist because they evolved step by step by natural selection. At each step, only those modifications that increased reproductive success were favored and retained in subsequent generations of offspring. Thus, every morphological feature and every behavior exists only because it was part of an adaptation that contributed to reproduction in ancestral populations. As a consequence, **mating systems** (the way animals find mates and care for offspring) play a crucial role in our understanding of primate societies.

Understanding the diverse reproductive strategies of nonhuman primates illuminates human evolution because we share many elements of our reproductive physiology with other species of primates.

To understand the evolution of primate mating systems, we must take into account that the reproductive strategies of living primates are influenced by their phylogenetic heritage as mammals. Mammals reproduce sexually. After conception, mammalian females carry their young internally. After they give birth, females suckle the young for an extended period of time. The mammalian male's role in the reproductive process is more variable than that of the female. In some species, males contribute little to their offspring's development besides a single sperm at the moment of conception, while in other species males defend territories, provide for their mates, and feed, carry, and protect their offspring.

Although mammalian physiology imposes some bounds on the nature of primate reproductive strategies, there is considerable room for diversity in primate mating systems and in reproductive behavior. Patterns of courtship, mate choice, and parental care vary greatly within the primate order. In some species, male reproductive success is determined mainly by success in competition with other males, while in others it is strongly influenced by female preferences. In many monogamous species, both males and females care for their offspring, while in most nonmonogamous species, females care for offspring and males compete with other males to inseminate females.

What aspects of mating do humans share with other primates? Until very recent times, all pregnant women nursed their offspring for an extended period, as with other primates. In nearly all traditional human societies, fathers contribute extensively to their children's welfare, providing resources, security, and social support. An understanding of the phylogenetic and ecological factors that shape the reproductive strategies of other primates may help us to understand how evolutionary forces shaped the reproductive strategies of our hominid ancestors, and the reproductive behavior of men and women in contemporary human societies.

The Language of Adaptive Explanations

In evolutionary biology, the term strategy *is used to refer to behavioral mechanisms that lead to particular courses of behavior in particular functional contexts, such as foraging or reproduction.*

Biologists often use the term **strategy** when describing certain aspects of the behavior of animals. For example, folivory is characterized as a foraging strategy, and monogamy is described as a mating strategy. When evolutionary biologists use the term, they mean something very different from what we normally mean when we use *strategy* to describe, say, a general's military maneuvers or a baseball manager's tactics. In common usage, *strategy* implies a conscious plan and a goal-directed course of action. Evolutionary biologists do not think other animals consciously decide to defend their territories, wean their offspring at a particular age, monitor their ingestion of secondary plant compounds, and so on. Instead, the term is used to refer to a set of behaviors occurring in a specific functional context, such as mating, parenting, or foraging. Strategies are the product of natural selection acting on individuals to shape the motivations, reactions, preferences, capacities, and choices that influence behavior. Predispositions that produce behaviors that led to greater reproductive success in ancestral populations have been favored by natural selection and represent adaptations.

For example, howler monkeys, who are folivorous, behave as though they know that some leaves are good for them in small quantities but harmful in large quantities and that young leaves contain more protein than older ones. Moreover, in different habitats they adjust the mix of plants in their diet in what appears to be a deliberate attempt to balance nutrients and to minimize toxins. But biologists doubt very much that howlers have any conscious knowledge of the nutritional content of the foods they eat or the components of an optimal diet. Instead, the underlying mechanisms that influence their decisions about what to eat, how much to eat, and what to avoid have been shaped by natural selection over many generations, producing the foraging behavior that we observe in nature.

The terms cost *and* benefit *refer to the effect of particular behavioral strategies on reproductive success.*

Different behaviors have different impacts on an animal's genetic fitness. Behaviors are said to be beneficial if they increase the genetic fitness of individuals, and costly if they reduce the genetic fitness of individuals. Thus, we argued in Chapter 6 that ranging behavior reflects a tradeoff between the benefits of exclusive access to a particular area and the costs of territorial defense. Reproductive success is the ultimate currency in which these tradeoffs are measured. Although this is a simple concept in principle, it is often very difficult to measure the costs and benefits associated with individual behavioral acts, particularly in long-lived animals like primates. Instead, researchers rely on indirect measures, such as foraging efficiency (measured as the quantity of nutrients obtained per unit time), and assume that, all other things being equal, behavioral strategies that increase foraging efficiency also enhance genetic fitness and will be favored by natural selection. We will encounter many other examples of this type of reasoning in the chapters that follow.

The Evolution of Reproductive Strategies

Primate females always provide extensive care for their young, while males do so in only a few species.

The amount of parental care varies greatly within the animal kingdom. In most species, parents do little for their offspring. For example, most frogs lay their eggs and never see their offspring again. In such species, the nutrients that females leave in the egg are the only form of parental care. In contrast, primates, like almost all birds and mammals and like some invertebrates and fish, provide much more than just the resources included in gametes. At least one parent, and sometimes both, shelter their young from the elements, protect them from predators, and provide them with food.

The *relative* amount of parental care provided by mothers and fathers also varies within the animal kingdom. In species without parental care, females produce large, nutrient-rich gametes, while males produce small gametes and supply only genes. Among species with parental care, however, all possible arrangements occur. Among primates, mothers always nurse their offspring and often provide extensive care

FIGURE 7.1

In all primate species, females nurse their young. In baboons, and many other species, females provide most of the direct care that infants receive.

(Figure 7.1). The behavior of fathers is much more variable. In most species, fathers give nothing to their offspring other than the genes contained in their sperm. However, in a minority of species, males are devoted fathers. In other taxa, patterns differ. For example, in most bird species, males and females form monogamous pairs and raise their young together (Figure 7.2), and there are even some bird species in which only males care for their chicks.

The amount of time, energy, and resources that the males and females of a species invest in their offspring has profound consequences for the evolution of virtually every aspect of that species' social behavior. As we shall see, species in which both parents care for their offspring are subject to radically different selection pressures than those species in which only one parent invests. We will also discover that species with equal parental investment differ from species with unequal parental care in morphology, behavior, and social organization. Thus, it is important to understand why the amounts and patterns of parental investment differ among species.

Males do not care for their offspring when 1) they can easily use their resources to acquire many additional matings or 2) when caring for their offspring would not appreciably increase the offspring's fitness.

FIGURE 7.2

In most species of birds, the male and female form a pair-bond and jointly raise their young. Here a bald eagle carries food to its hungry brood.

At first glance, it seems odd that most primate males fail to provide any care for their offspring. Surely, if the males helped their mates, they would increase the chances of their offspring surviving to adulthood. Therefore, we might expect paternal care to be favored by natural selection (Figure 7.3).

If time, energy, and other resources were unlimited, this reasoning would be correct. But in real life, time, energy, and material resources are always in short supply. The effort that an individual devotes to caring for offspring is effort that cannot be spent competing for prospective mates. Natural selection will favor individuals that allocate effort among these competing demands so as to maximize the number of surviving offspring that they produce.

To understand the evolution of asymmetries in parental investment, we must identify the conditions under which one sex can profitably reduce its parental effort at the expense of its partner. Consider a species in which most males help their mates feed and care for their offspring. There are a few males, however, with a heritable tendency to invest less in their offspring. We will refer to these two types as "investing" and "noninvesting" fathers. Since time, energy, and resources are always limited, males that devote more effort to caring for offspring must allocate less effort to competing for access to females. On average, the offspring of noninvesting males will receive less care than will the offspring of investing males, making them less likely to survive and to reproduce successfully when they mature. On the other hand, since noninvesting males are not kept busy caring for their offspring, they will on average acquire more mates than will investing males. Mutations favoring the tendency to provide less parental care will increase in frequency when the benefits to males, measured in terms of the increase in fitness gained from additional matings, outweigh the costs to males, measured as the decrease in offspring fitness due to a reduction in paternal care.

This reasoning suggests that unequal parental investment will be favored when one or both of the following are true:

1. Acquiring additional mates is relatively easy, so considerable gains are achieved by allocating additional effort to attracting mates.
2. The fitness of offspring raised by only one parent is high, so the payoff for additional parental investment is relatively low.

Thus, the key factors are the costs of finding additional mates and the benefits associated with incremental increases in the amount of care that offspring receive. When females are widely separated, for example, it may be difficult for males to locate them. In these cases, males may profit more from helping their current mates and investing in their offspring than from searching for additional mates. If females are capable of rearing their offspring alone and need little help from males, then investing males may be at a reproductive disadvantage compared with males that abandon females after mating and devote their efforts to finding eligible females.

The mammalian reproductive system commits primate females to invest in their offspring.

So far, there is nothing in our reasoning that says that if only one sex invests, it should always be females. Why aren't there primate species in which males do all the work, and females compete with each other for access to males? This is not simply a theoretical possibility—female sea horses deposit their eggs in their mate's brood pouch and then swim away and look for a new mate (Figure 7.4). There are whole families of fish in which male parental care is more common than female parental care; and in several species of birds, including rheas, spotted sandpipers, and jacanas, females abandon their clutches after the eggs are laid, leaving their mates to feed and protect the young.

In primates and other mammals, the bias toward female investment arises because females lactate and males do not. Among primates, pregnancy and lactation commit mammalian females to invest in their young, and limit the benefits of male investment in offspring. Since offspring depend on their mothers for nourishment during pregnancy and after birth, females cannot abandon their young without greatly reducing the offspring's chances of surviving. On the other hand, males are never capable of rearing their offspring without help from females. Therefore, when only one sex invests in offspring, it is invariably females that do so. Sometimes males can help females by defending territories or by carrying infants so that the mother can feed more efficiently, as siamangs and owl monkeys do. In most cases, however, these benefits are relatively insignificant, and selection favors males that allocate more time and energy to mating than to caring for their offspring.

You may be wondering why selection has not produced males that are able to lactate. As we noted in Chapter 3, most biologists believe the developmental changes that would enable males to lactate would also make them sterile. This is an example of a developmental constraint.

Reproductive Strategies of Females

So far, we have established that female primates often invest more heavily in their offspring than do males. The next step is to consider the reproductive strategies of primate females in more detail.

FIGURE 7.3

In most nonmonogamous species, males have relatively little contact with infants. Although males, like this bonnet macaque, are sometimes quite tolerant of infants, they rarely carry, groom, feed, or play with infants. (Photograph by Kathy West.)

FIGURE 7.4

In sea horses, males carry fertilized eggs in a special pouch and provide care for their young as they grow.

Female primates invest heavily in each of their offspring.

Pregnancy and lactation are time-consuming and energetically expensive activities for female primates. The duration of pregnancy plus lactation ranges from 102 days in the tiny mouse lemur to 1839 days in the hefty gorilla. In primates, as in most other animals, larger animals tend to have longer pregnancies than do smaller animals, but primates have considerably longer pregnancies than we would expect based on their body sizes alone. The extended duration of pregnancy in primates is related to the fact that brain tissue develops very slowly. Primates have very large brains in relation to their body sizes, so additional time is needed for fetal brain growth and development during pregnancy. Primates also have an extended period of dependence after birth, further increasing the amount of care mothers must provide. Throughout this period, mothers must meet not only their own nutritional requirements but also those of their growing infants. In some species, offspring may weigh as much as 30% of their mother's body weight at the time of weaning.

The energy costs of pregnancy and lactation impose important constraints on female reproductive behavior. Since it takes so much time and energy to produce an infant, each female can rear a relatively small number of surviving infants during her lifetime (Figure 7.5). For example, a female toque macaque who gives birth for the first time when she is five years old and survives to the age of 20 would produce 15 infants if she gave birth annually and all of her infants survived. This number undoubtedly represents an upper limit; in the wild a substantial fraction of all toque macaque infants die before they reach reproductive age, intervals between successive live births often last two years, and some females die before they reach old age. Thus, most toque macaque females will produce a relatively small number of surviving infants over the course of a lifetime, and each infant represents a substantial proportion of a female's lifetime fitness. Therefore, we would expect mothers to be strongly committed to the welfare of their young.

FIGURE 7.5

A female bonnet macaque produced twins, but only one survived beyond infancy. Twins are common among marmosets and tamarins but are otherwise uncommon among monkeys and apes. (Photograph by Kathy West.)

A female's reproductive success depends largely on her ability to obtain enough resources to support herself and her offspring.

In most species of primates, including humans, females must achieve a minimum nutritional level in order to ovulate and to conceive. This means that females may sometimes be unable to conceive because their nutritional status is poor, they may have long periods of infertility while they recover from the rigors of their last pregnancy, and they may not always be able to nourish themselves or their newborns adequately.

For animals living in the wild, without take-out pizza or 24-hour grocery stores, getting enough to eat each day is often a serious dilemma. There is considerable evidence that female reproductive success is limited by the availability of resources within their local habitat. For example, there are a number of sites in Japan where free-ranging monkeys' natural diets have been supplemented with wheat, sweet potatoes, rice, and other foods by humans for many years (Figure 7.6). When their diet is sup-

FIGURE 7.6

At a number of locations in Japan, indigenous monkeys are fed regularly. The size of these artificially fed groups has risen rapidly, indicating that population growth is limited by the availability of resources.

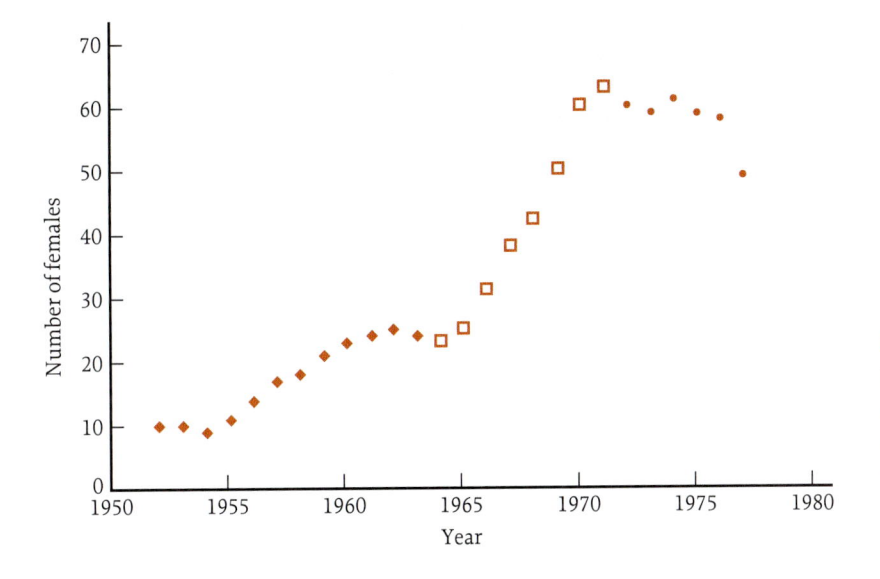

FIGURE 7.7

The size of the Koshima troop of Japanese monkeys grew rapidly when they were provisioned intensively (□), and then dropped when provisioning was restricted (•).

plemented in this way, females grow faster, mature earlier, survive longer, and produce infants at shorter intervals. Figure 7.7 shows changes in the size of one such group in Japan before, during, and after a period of intense supplementation. Similar effects have been documented under natural conditions. For example, a long period of environmental deterioration led to a drastic decline in the size and number of baboon groups in Amboseli, Kenya. In 1963 and 1964, biologists Jeanne and Stuart Altmann (now at Princeton University) counted 2500 baboons in 51 groups, but in 1979 there were only 123 baboons in five groups. During this period, female birth rates declined and mortality among immature animals increased.

Sources of Variation in Female Reproductive Performance

High-ranking females tend to reproduce more successfully than do low-ranking females.

As we explained in Chapter 6, females often compete for access to food resources that they need in order to reproduce successfully. In some situations, females form dominance hierarchies, and high-ranking females have priority of access to valued foods (Figure 7.8). If dominance rank influences access to valuable resources and access to resources influences female reproductive success, then we should expect to find a positive correlation between dominance rank and reproductive success. Data from a wide range of species indicate that high rank confers reproductive advantages on females.

In multimale, multifemale groups of Old World monkeys that form quite stable matrilineal dominance hierarchies, female dominance rank is correlated with various aspects of females' reproductive performance. For example, in Amboseli, Kenya, the offspring of high-ranking female baboons grow faster and mature earlier than do the offspring of low-ranking females. In captive vervet groups, high-ranking females have shorter interbirth intervals than lower ranking

FIGURE 7.8

Female primates sometimes fight over access to food and other resources.

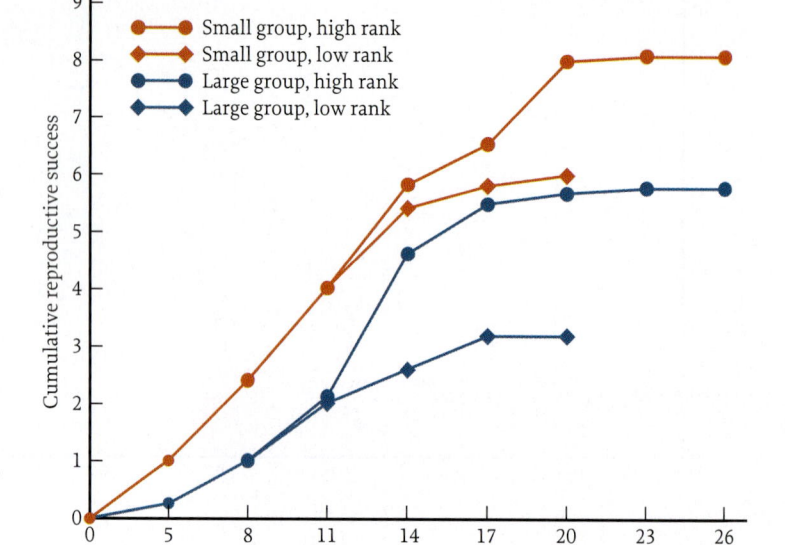

FIGURE 7.9

In free-ranging groups of long-tailed macaques, both group size and dominance rank influence females' lifetime reproductive success. In general, females living in smaller groups reproduce more successfully than do females living in larger groups. But in both large and small groups, high-ranking females reproduce more successfully than low-ranking females.

females do. In some macaque populations the offspring of high-ranking females are more likely to survive to reproductive age than are the offspring of lower-ranking females. Associations between dominance rank and reproductive success may produce substantial variation in lifetime fitness among females, particularly if females maintain the same rank over the course of their lives, as female macaques and baboons typically do. Thus, Maria van Noordwijk and Carel van Schaik of Duke University have found substantial differences in the lifetime reproductive success of high-, middle-, and low-ranking long-tailed macaques (Figure 7.9).

Marmoset and tamarin groups in the wild often contain more than one adult female, but the dominant female is usually the only one that breeds successfully. Reproductive activity of subordinate females is suppressed in the presence of dominant females, as subordinate females do not cycle normally. When subordinate females do breed, their infants may be killed by dominant females that have infants of their own.

Anne Pusey and her colleagues at the University of Minnesota have shown that the offspring of high-ranking female chimpanzees are more likely to survive to the age of weaning than are the offspring of low-ranking females. In addition, their daughters grow faster and mature earlier than do the daughters of low-ranking females. These differences create substantial differences in lifetime fitness for high- and low-ranking female chimpanzees at Gombe Stream National Park, Tanzania (Figure 7.10).

Red howlers live in groups that contain only two to four adult females and one or two adult males. Long-term studies of red howlers by Teresa Pope of Duke University and Carolyn Crockett of the Washington Regional Primate Center have shown that in sparsely populated habitats, new groups are formed when unrelated females meet, form social bonds, attract males, and establish territories. As habitats become more crowded, it becomes harder and harder to establish new territories and female dispersal becomes more costly. However, not all females can remain in their natal groups because group size is confined within narrow limits. This generates intense competition among females over recruitment opportunities for their daughters, and in most cases only the daughters of dominant females are able to remain in their natal groups.

In Hanuman langurs, female rank also influences female reproductive performance. In this species, female rank is inversely related to age (Figure 7.11). Long-term

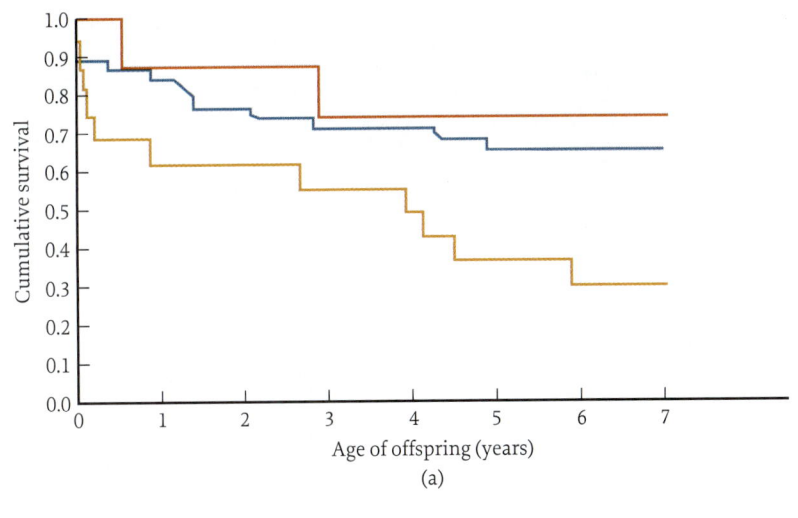

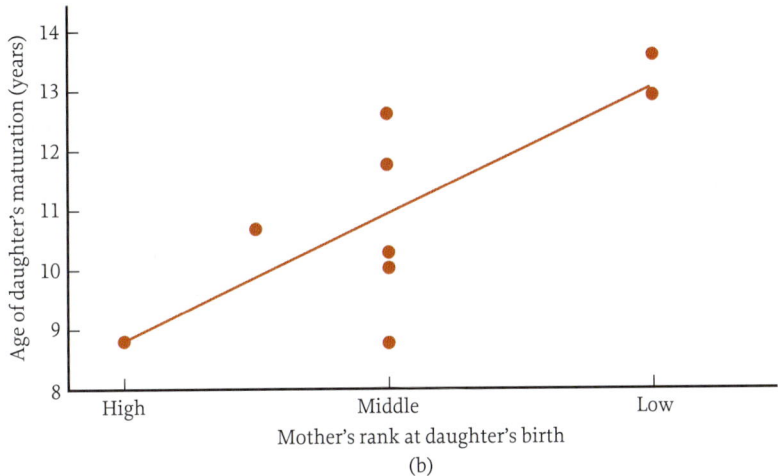

REPRODUCTIVE STRATEGIES OF FEMALES

FIGURE 7.10

Among Chimpanzees at Gombe Stream National Park, Tanzania, female rank influences reproductive performance. (a) The offspring of high- (red line) and middle-ranking (purple line) females are more likely to survive to weaning age than are the offspring of low-ranking females (yellow line). (b) Daughters of high-ranking females also mature at earlier ages than do the daughters of lower-ranking females.

FIGURE 7.11

(a) A female Hanuman langur at Jodphur threatens another group member. (b) Among Hanuman langur females, dominance rank is inversely related to age, so the youngest females hold the highest ranks and their rank declines as they age. (a, Photograph courtesy of Carola Borries.)

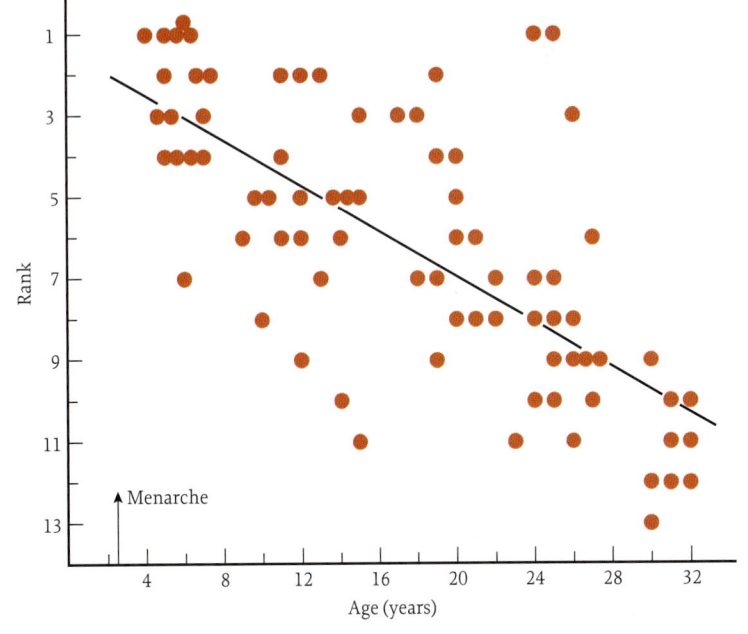

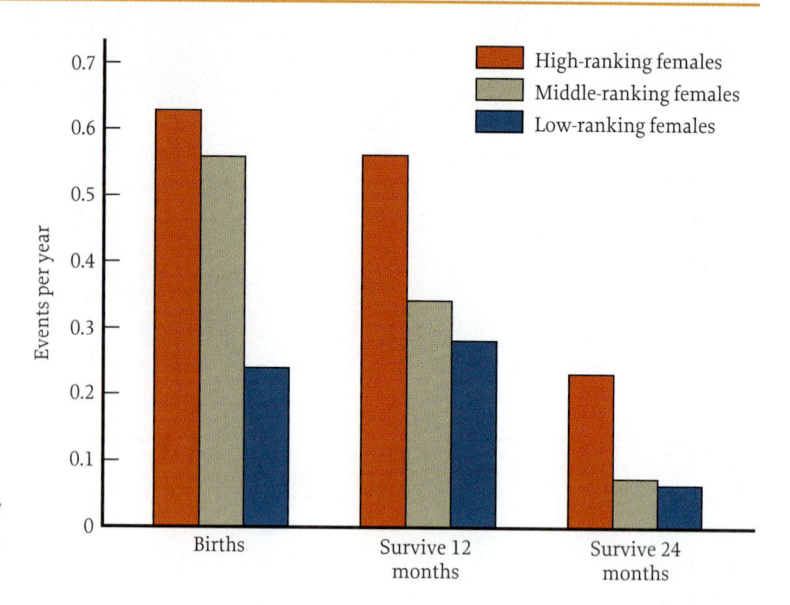

FIGURE 7.12

Female Hanuman langurs reproduce more successfully when they are young and hold high ranks than when they are older and have a lower rank.

studies of Hanuman langurs near Jodphur, India, conducted by a group of German primatologists, including Carola Borries now at the State University of New York at Stony Brook and Volker Sommer now at University College London, have shown that young, high-ranking females reproduce more successfully than do older, lower-ranking females (Figure 7.12). Studies of Hanuman langurs at Ramnagar in southern Nepal conducted by another group of German primatologists, including Andreas Koenig, Carola Borries, and Paul Winkler, have found that high-ranking females manage to commandeer higher-quality food patches and are consequently able to maintain higher levels of body fat. Females in good condition have higher fertility rates than do females in poor condition.

Some researchers question the importance of the effects of dominance rank on reproductive success because significant effects of dominance rank are not found in every study. Among baboons at Gombe, for example, high-ranking female baboons suffered higher rates of miscarriages and reproductive failure than did lower-ranking females, resulting in lower lifetime fitness for the highest-ranking females. However, positive correlations between rank and reproductive success greatly outnumber negative correlations. There is little evidence that it is ever advantageous to be low ranking.

REPRODUCTIVE TRADEOFFS

Females must make a tradeoff between the number of offspring they produce and the quality of care that they provide.

Just as both males and females must allocate limited effort to parental investment and mating, females must apportion resources among their offspring. All other things being equal, natural selection will favor individuals that are able to convert effort into offspring most efficiently. Since mothers have a finite amount of effort to devote to offspring, they cannot maximize both the quality and the quantity of the offspring they produce. If a mother invests great effort in one infant, she must reduce her investment in others. If a mother produces many offspring, she will be unable to invest very much in any of them.

REPRODUCTIVE STRATEGIES OF FEMALES

In nature, maternal behavior reflects this tradeoff when a mother modifies her investment in relation to an offspring's needs. Initially, infants spend virtually all of their time in contact with their mothers. The very young infant is entirely dependent on its mother for food and transportation and is unable to anticipate or to cope with environmental hazards. At this stage, mothers actively maintain close contact with their infants (Figure 7.13): retrieving them when they stray too far, signaling to them when they are about to depart, and scooping them up when danger arises.

As infants grow older, however, they become progressively more independent and more competent. They venture away from their mothers to play with other infants and to explore their surroundings. They begin to sample food plants, sometimes mooching scraps of their mother's food. They become aware of the dangers around them, attending to alarm calls given by other group members and reacting to disturbances within the group. Mothers use a variety of tactics to actively encourage their infants to become more independent. Initially, they may subtly resist their infants' attempts to suckle. At the same time, mothers may begin to encourage their infants to travel independently by moving off before the infant has climbed onto their backs, by shrugging their infants off their backs while moving, or by simply failing to retrieve their infants when ready to depart. Although infants may protest these developments with tantrums or whimpers, and persistently attempt to ride on their mothers and to nurse, they eventually become independent of their mothers. Nursing is gradually limited to brief and widely spaced bouts that may provide the infant with more psychological comfort than physical nourishment. At this stage, infants are carried only when they are ill, injured, or in great danger.

FIGURE 7.13

A female chimpanzee sits beside her youngest infant in Gombe Stream National Park in Tanzania.

The changes in maternal behavior reflect the shifting balance between the requirements of the growing infant and the energy costs to the mother of catering to her infant's needs. As infants grow older, they become heavier to carry and require more food, which imposes substantial burdens on mothers (Figure 7.14). As infants become more capable of feeding themselves and of traveling independently, mothers can gradually limit investment in their older infants without jeopardizing the infants' welfare. This enables mothers to conserve resources that can be allocated to subsequent infants. Moreover, since lactation inhibits ovulation in many primate species, mothers must wean their present infant before they are able to conceive another. Phyllis Lee of Cambridge University and her colleagues have discovered that although the duration of lactation varies within and among species, generally primate infants are weaned when they reach about four times their birth weight (Figures 7.15 and 7.16).

FIGURE 7.14

As infants become older, they become a bigger burden on their mothers. Here, a female baboon leaps across a ditch carrying her infant on her back. Later, the mother will encourage her infant to leap across the ditch on her own.

Male reproductive behavior in primates is strongly affected by the distribution of females.

Earlier we saw that male reproductive strategies are influenced by two factors: the costs of obtaining additional matings, and the benefits to infants of receiving care from both parents rather than from just one. We have established that primate females are generally prepared to raise their offspring with relatively little assistance from males. Thus, male mating strategies are likely to be determined largely by the

FIGURE 7.15

Primate infants are generally weaned when they reach about four times their birth weight, although the duration of lactation and infant growth rates vary considerably. The threshold weaning weight probably reflects constraints on the mother's ability to meet the energetic demands of growing infants. Each point represents the average value of birth weight and weaning weight for a given primate genus. ○ represents a human foraging group, the !Kung.

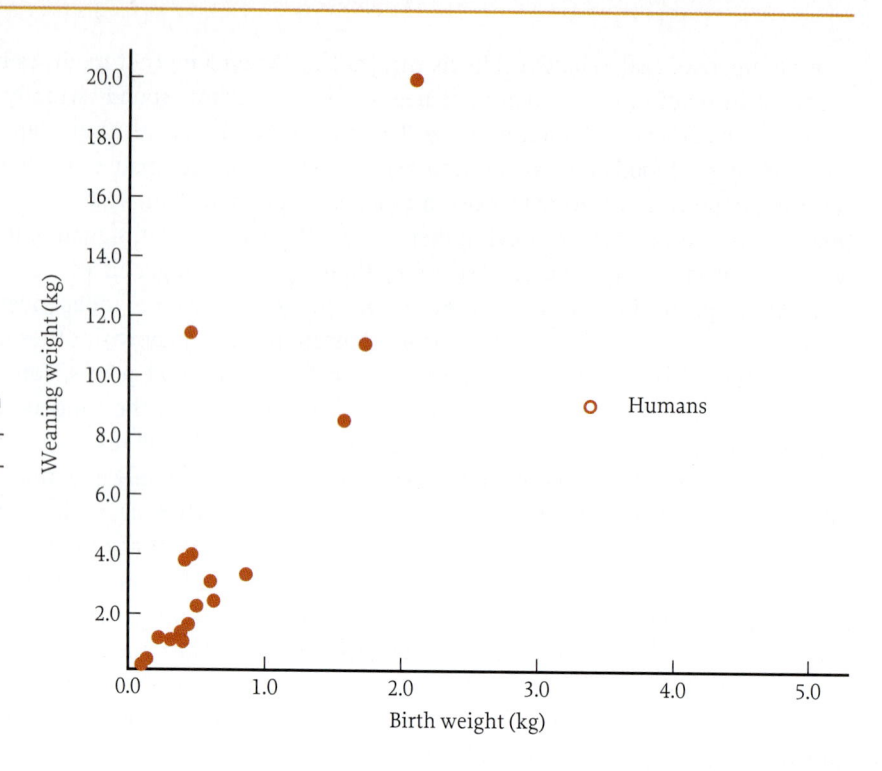

FIGURE 7.16

A large infant squirrel monkey is being carried by its mother. By the time they are weaned, infants in some species may weigh up to a third of their mother's weight. (Photograph by Sue Boinski.)

costs of obtaining additional matings, and these costs will in turn depend mainly on the distribution of females. If females are sparsely dispersed in the environment, then it may be difficult and time-consuming for males to locate prospective mates. In such cases, males may benefit more from establishing a pair-bond with a single female and investing in their offspring than from seeking additional matings with a number of females. If females are clumped together in groups, however, then the cost of finding mates is likely to be low. Under those circumstances, males are expected to focus on attempting to obtain additional matings, rather than investing in their offspring.

Sexual Selection and Male Mating Strategies

Sexual selection leads to adaptations that allow males to compete more effectively with other males for access to females.

So far, we have seen that primate females invest heavily in each of their young and produce relatively few offspring over the course of their lives. Moreover, most primate females can raise their offspring without help from males. Female reproductive success is limited by access to food, not access to mates (Figure 7.17). Males can potentially produce progeny from many females, and as a result males compete for access to females. Characteristics that increase male success in competition for mates will spread as a result of what Darwin called sexual selection.

It is important to understand the distinction between natural selection and sexual selection. Most kinds of natural selection favor phenotypes in both males and females that enhance their ability to survive and reproduce. Many of these traits are related to resource acquisition, predator avoidance, and offspring care. **Sexual selection** is a special category of natural selection that favors traits that increase success in competition for mates, and will be expressed most strongly in the sex whose access to members of the opposite sex is most limited. Sexual selection may favor traits like the peacock's tail, red deer's antlers, and the hamadryas baboon's mane, even if those traits reduce the ability of the animal to survive or acquire resources, outcomes not usually favored by natural selection (Figure 7.18).

Sexual selection is often much stronger than ordinary natural selection.

In mammalian males, sexual selection can have a greater effect on behavior and morphology than other forms of natural selection do. This is because male reproductive success usually varies much more than female reproductive success. Data from long-term studies of lions conducted by Craig Packer and Anne Pusey of the University of Minnesota demonstrate that the lifetime reproductive success of the most successful males is often much greater than that of even the most successful females (Figure 7.19). The same pattern is likely to hold for nonmonogamous primates. A primate male who succeeds in competition may sire scores of offspring, while a successful female might give birth to five or 10 offspring. Unsuccessful males and females will fail to reproduce at all. Since the strength of selection depends on how much variation in fitness there is among individuals, sexual selection acting on male primates can be much stronger than selective forces acting on female primates. (As an aside, in species like sea horses, in which males invest in offspring and females do not, the entire pattern is reversed. Sexual selection acts much more strongly on females than on males.)

FIGURE 7.17
For female primates, access to resources has a bigger impact on reproduction than access to potential mates does.

There are two types of sexual selection: 1) intrasexual selection results from competition among males, and 2) intersexual selection results from female choice.

Many students of animal behavior subdivide sexual selection into two categories: in species in which females can choose whom they mate with, selection favors traits

(a)   (b)

FIGURE 7.18
Sexual selection can favor traits not favored by natural selection. (a) The peacock's tail hinders his ability to escape from predators, but it enhances his attractiveness to females. Female peahens are attracted to males with the most eyespots in their trains. (b) Male red deer use their antlers when they fight with other males. Red deer antlers are a good example of a trait that has been favored by sexual selection.

FIGURE 7.19

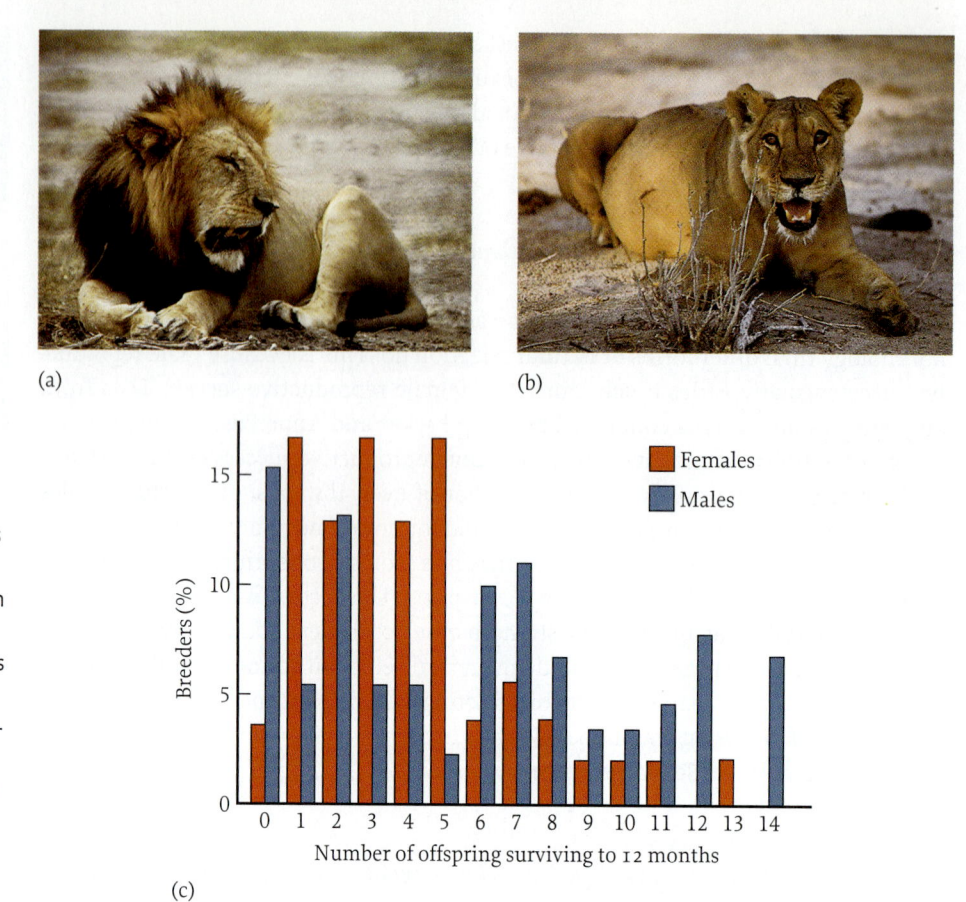

(a) The reproductive success of male lions is considerably more variable than (b) the reproductive success of female lions. (c) In Serengeti National Park and the Ngorongoro Crater of Tanzania, few female lions fail to produce any surviving cubs, but most females produce less than six surviving cubs over the course of their lives. Many males fail to produce any cubs, and a few males produce many cubs.

that make males more attractive to females. This is **intersexual selection.** In species where females cannot choose their mates, access to females will be determined by competition among males. In such species, **intrasexual selection** favors traits that enhance success in male-male competition.

INTRASEXUAL SELECTION

Competition among males for access to females favors large body size, large canine teeth, and other weapons that enhance male competitive ability.

For primates and most other mammals, intrasexual competition is most intense among males. Male-male competition can take different forms, but the simplest occurs when males drive other males away from females. Males that regularly win such fights have higher reproductive success than those that lose. Thus, intrasexual selection favors features such as large body size, horns, tusks, antlers, and large canine teeth that enable males to be effective fighters. For example, male gorillas compete fiercely over access to groups of females, and males weigh twice as much as females and have longer canine teeth.

As explained in Chapter 5, when the two sexes consistently differ in size or appearance, they are said to be sexually dimorphic (Figure 7.20). The body sizes of males and females represent compromises among many competing selective pressures. Larger animals are better fighters and are less vulnerable to predation, but they also need more food and take longer to mature. Intrasexual competition favors larger body size,

FIGURE 7.20

Adult male baboons are nearly twice the size of adult females. The degree of sexual dimorphism in body size is most pronounced in species with the greatest competition among males over access to females.

larger teeth, and other traits that enhance fighting ability. Males compete over females, while females compete over resources but generally do not compete over mates. However, the effect of intrasexual competition among males is quantitatively greater than the effect of competition among females because the fitness payoff to a successful male is greater than it is to a successful female. Therefore, sexual selection is much more intense than ordinary natural selection. As a result, intrasexual selection leads to the evolution of sexual dimorphism.

The fact that sexual dimorphism is greater in primate species forming one-male and multimale groups than in monogamous species indicates that intrasexual selection is the likely cause of sexual dimorphism in primates.

If sexual dimorphism among primates is the product of intrasexual competition among males over access to females, then we should expect to see the most pronounced sexual dimorphism in the species in which males compete most actively over access to females. One indirect way to assess the potential extent of competition among males is to consider the ratio of males to females in social groups. In general, male competition is expected to be most intense in social groups where males are most outnumbered by females. This prediction might seem paradoxical at first, since we might expect to have more competition when there are more males present. The key to resolving this paradox is to remember that there are approximately equal numbers of males and females at birth in most natural populations. In species that form one-male groups, there are many **bachelor males** (males who don't belong to social groups) who exert constant pressure on resident males. In species that form monogamous pair-bonds, each male is paired with a single female, reducing the intensity of competition among males over access to females.

Comparative analyses conducted by Paul Harvey of Oxford University and Tim Clutton-Brock of Cambridge University have demonstrated that the extent of sexual dimorphism in primates corresponds roughly to the form of social groups in which the males live (Figure 7.21). There is little difference in body weight or canine size between males and females in species that typically form monogamous groups, such as gibbons, titi monkeys, and marmosets. At the other extreme, the most pronounced di-

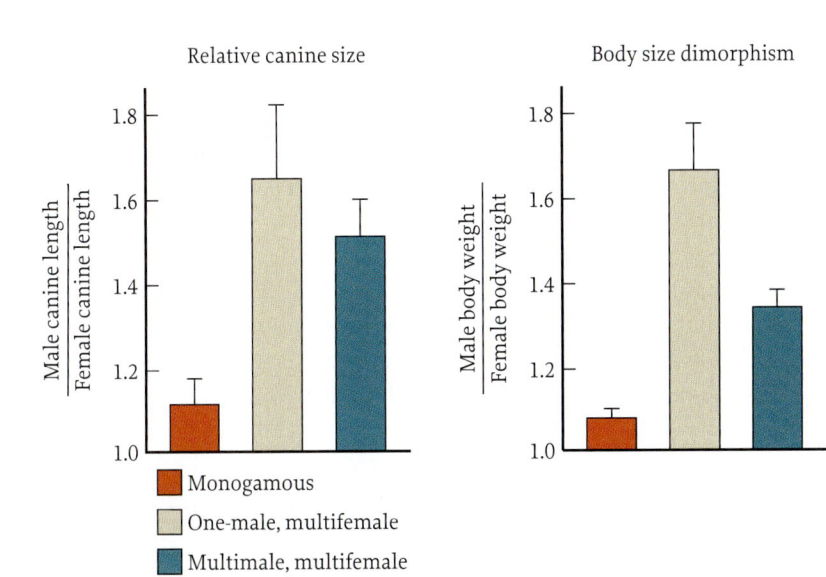

FIGURE 7.21

The degree of sexual dimorphism is a function of the ratio of males to females in social groups. Relative canine size (male canine length divided by female canine length) and body size dimorphism (male body weight divided by female body weight) are greater in species that form one-male, multifemale groups than in species that form multimale, multifemale groups or monogamous groups.

morphism is found in species, such as gorillas and black-and-white colobus monkeys, that live in one-male, multifemale groups. And in species that form multimale, multifemale groups, the extent of sexual dimorphism is generally intermediate between these extremes. Thus, sexual dimorphism is most pronounced in the species in which the ratio of males to females living in bisexual groups is lowest (that is, with the highest relative number of females).

> *In multimale, multifemale groups, where females mate with several males during a given estrous period, sexual selection favors increased sperm production.*

In most primate species, as with mammals in general, the female is receptive to mating mainly during the portion of her reproductive cycle when fertilization is possible. That period of time is called **estrus.** In primate species that live in multimale, multifemale groups, females can potentially mate with several different males during a single estrous period. In such species, sexual selection favors increased sperm production because males that deposit the largest volume of sperm in the female reproductive tract have the greatest chance of impregnating them. Competition in the quantity of sperm is likely to be relatively unimportant in monogamous species, because females mainly mate with their own partners. Since sperm production involves some cost to males, monogamous males may do better by guarding their partners when they are sexually receptive than by producing large quantities of sperm. Similarly, competition in sperm quantity probably does not play an important role in species that form one-male, multifemale groups. In these species, competition among males is over access to groups of females, which favors traits related to fighting ability. If resident males are able to exclude other males from associating with females in their groups, there may be little need to produce additional quantities of sperm.

Social organization is associated with testes' size, much as we would expect. Males with larger testes typically produce more sperm than do males with smaller testes, and males that live in multimale groups have much larger testes in relation to their body size than males that live in either monogamous or one-male, multifemale groups (Figure 7.22). Interestingly, in polyandrous groups where a single breeding female mates regularly with multiple males, males have relatively large testes.

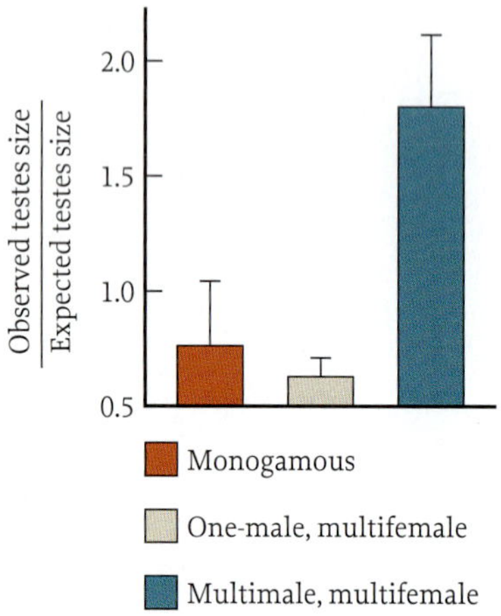

FIGURE 7.22

The average size of testes in species that typically form monogamous and one-male, multifemale groups is relatively smaller than the average size of testes in multimale, multifemale groups. Here observed testes' weight is divided by the expected testes' weight to produce relative testes' size. The expected testes' weight is derived from analyses that correct for the effects of body size.

INTERSEXUAL SELECTION

> *Intersexual selection favors three kinds of traits among males: 1) those that increase the fitness of their mates, 2) those that indicate good genes and thus increase the fitness of the offspring, and 3) nonadaptive traits that make males more conspicuous to females.*

Darwin also realized that females might be attracted to males that exhibit particular traits. In this case, intersexual selection favors males that are pleasing to females rather than males that can defeat other males. Modern sexual selection theory now identifies three basic modes of female choice. First, if males with certain traits confer direct benefits on females, then selection will favor females that mate selectively with males possessing such traits. This will increase the frequency of such traits among males. Thus, if females mate preferentially with males that defend superior territories, protect offspring more vigorously, provision the young better, or confer more material benefits on females than other males do, these male traits will be favored.

Second, sexual selection may also favor female preferences if females are able in some way to distinguish male genetic quality and choose mates with desirable genes. These females are at an advantage because their offspring will carry certain genes that give them a greater chance of surviving and reproducing successfully. The gaudy peacock and the noisy sage grouse may be examples of such preferences. Males gather together in a small area called a lek, and each male displays his tail feathers to attract female attention (Figure 7.23). Marion Petrie of Oxford University has found that female peahens never mate with the first male they encounter on the lek, and show distinct preferences for the males with the most eyespots on their tail feathers. The offspring of the most richly ornamented males grow faster and survive better than the offspring of other males. Thus, female mate preferences enhance offspring fitness. Again, it's important to remember that we need not imagine that peahens, or any other animals, assess male genetic quality *consciously*. Instead, selection may have favored preferences for specific traits that are reliably associated with genetic quality. By choosing males with the most eyespots, hens may be inadvertently and unconsciously choosing males with desirable genotypes.

Third, females may prefer males that exhibit distinctive traits (such as conspicuous coloration, exaggerated morphological characters, or elaborate courtship behaviors) for nonadaptive reasons. Even if such characters do not increase male fitness directly, they may be favored if females can discriminate and consistently select mates that most often display preferred traits. For example, male frogs call to attract mates, and the sensory system of female frogs seems to respond more strongly to some tones in male calls than to others. Researchers have discovered that male frogs whose calls are most readily heard attract the most mates.

Surprisingly, it turns out that female choice can also lead to the exaggeration of female preferences themselves. To see why, suppose there are two types of females: very picky females who prefer males with only the most exaggerated male trait—say, the deepest tones of a male frog's call or the longest tails of a male bird—and less picky females who will mate with a wider range of males. It is easy to see that the existence of more picky females will increase the reproductive success of males with the most exaggerated traits, and therefore increase the degree of exaggeration of the male trait in the population. However, the same process can also lead to the spread of more picky females. This is because the offspring of more picky females will tend to carry both the genes for pickiness and the genes for the exaggerated male trait, creating a positive correlation between the female preference and the exaggerated male trait (see Chapter 3). As a result of the correlated response to selection, the female choice that increases the reproductive success of exaggerated males will also increase the pickiness of females. If conditions are right, this can lead to a runaway process in which females become pickier and pickier about a male trait, and male traits become more and more exaggerated, even if they impair the male's ability to survive.

There are only a few morphological traits among primates that seem to have evolved because they help males to attract females.

The best candidates for traits that help primate males to attract females are found among mandrills and proboscis monkeys. Both the face and the penis of the male mandrill are brilliantly colored. Their faces are striped red, white, and blue like an exotic mask (Figure 7.24), while the female mandrill's face is duller and less conspicuous.

FIGURE 7.23

Male sage grouse congregate on leks and display to attract females. (Photograph courtesy of Mark Chappell.)

FIGURE 7.24

Male mandrills have brightly colored faces. This trait may be favored by intersexual selection, but it is not known whether females actually prefer to mate with brightly colored males.

FIGURE 7.25
Male proboscis monkeys have long pendulous noses, while females have much smaller upturned noses. This trait may be favored by intersexual selection, although we do not know whether females are attracted to males with long noses.

The proboscis monkey is named for its oddly shaped nose: males have quite long, pendulous noses, while females have much smaller noses that turn up at the end (Figure 7.25). There is pronounced sexual dimorphism in body size among both these species as well. Very little is known about mandrills, who live in dense primary rain forests, or about proboscis monkeys, who live in mangrove swamps, so we don't really know whether females actually prefer mandrill males with brightly colored faces or proboscis males with elongated noses. We also do not know which of the three modes of female choice discussed earlier may be operating. However, since these traits do not seem to be directly related to male-male competition and are more fully developed in males than in females, it seems possible these traits evolved in response to female preferences.

Male Reproductive Tactics

Morphological evidence suggests that in monogamous species, male-male competition is relatively muted, while in nonmonogamous species male-male competition is more intense. As we will see in the remainder of this chapter, sexual selection has shaped male mating strategies as well as their morphology.

INVESTING MALES

Monogamous pair-bonding is generally associated with high levels of paternal investment.

In species that form monogamous pair-bonds, males do not compete directly over access to females. In these species, males' reproductive success depends on their ability to establish territories, find mates, and rear surviving offspring. In pair-bonded species, mate guarding and offspring care are important components of males' reproductive tactics.

Mate guarding may be an important component of pair-bonded males' reproductive effort, as monogamy doesn't necessarily imply fidelity. Numerous genetic studies of supposedly monogamous birds have demonstrated that a large fraction of the young are not sired by the female's mate. Pair-bonded titi monkeys and gibbons have also been seen copulating with members of neighboring groups. If females occasionally participate in extra-pair copulations, their partners may benefit from keeping close watch on them. Ryne Palombit of Rutgers University, who has studied the dynamics of pair-bonding in gibbons, suspects that males do just that. Males are principally responsible for maintaining proximity to their female partners, and most males groom their mates more than they are groomed in return (Figure 7.26).

Pair-bonded males tend to invest heavily in their mates' offspring. In titi monkeys and owl monkeys, which have been studied in the forests of Peru by Patricia Wright of the State University of New York at Stony Brook, adult males play an active role in caring for infants. They carry them much of the time, share food with them, groom them, and protect them from predators. Male siamang are also helpful fathers—carrying their infants for long periods every day.

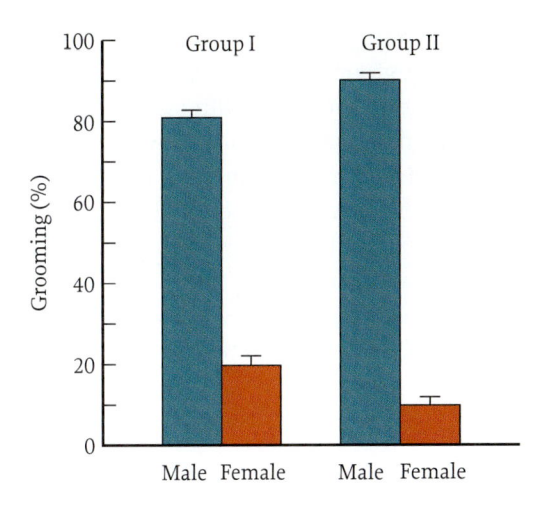

MALE REPRODUCTIVE TACTICS

FIGURE 7.26
In white-handed gibbon groups, males groom their mates far more than they are groomed in return. Males' solicitous attention to females may be a form of mate guarding.

In polyandrous species males invest heavily in offspring, but the reproductive benefits are not clear.

Polyandry presents a puzzle for evolutionary biologists because males who share access to a single female can produce only half as many offspring as males who monopolize access to a female. Why do males put up with this arrangement? In some cases, such as the Tasmanian hen, polyandrous males are brothers and have a common genetic interest in the offspring of their common mate. (We will explain how kinship influences behavioral strategies in more detail in the next chapter.) In other cases, males may temporarily accept a subordinate role in hopes of eventually replacing the dominant male. Finally, it is possible that polyandry may benefit males if their joint investment in offspring increases female fertility. The rarity of polyandry in nature suggests that the costs of sharing access to females generally outweigh the benefits.

In primates, there are very few examples of polyandry. Marmoset and tamarin groups are sometimes classified as polyandrous because some groups contain one adult female and two adult males (Figure 7.27). However, the social organization in these species is quite variable: some groups contain one adult female and two adult males, some contain a single adult male and adult female, and others contain multiple adult males and females.

Behavioral and genetic data suggest that reproductive benefits are not divided equally among males in these species. In most species of marmosets and tamarins, the dominant male monopolizes matings with receptive females. These behavioral data are consistent with very limited evidence from genetic analyses. Leslie Digby of Duke University and her colleagues have analyzed the paternity of infants born in three free-ranging groups of common marmosets in Brazil. These three groups each contained two adult males and one or two breeding females. They found that in each group the dominant male fathered nearly all of the infants.

The presence of multiple males does seem to enhance female fertility. Marmosets and tamarins are unusual among primates because they usually produce twins, and females

FIGURE 7.27
Some callitrichid groups contain more than one adult male and a single breeding female. In at least some groups, mating activity is limited to the dominant male in the group, even though all the males participate in the care of offspring.

FIGURE 7.28

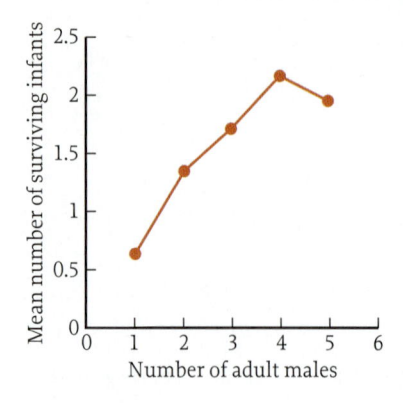

In tamarin groups, males clearly contribute to females' reproductive success. Tamarin groups that contain several adult males produce more surviving infants than do groups that contain only one adult male.

produce litters at relatively short intervals, sometimes twice a year. Males play an active role in child care, frequently carrying infants, grooming them, and sharing food with them. Data compiled by Paul Garber of the University of Illinois at Urbana-Champaign show that groups with multiple adult males reared more surviving infants than did groups with only one male (Figure 7.28). In contrast, groups with multiple females produced slightly fewer infants than did groups with only one female resident.

Thus, while females seem to benefit from having multiple males to assist them in rearing offspring, the reproductive payoffs for males are harder to document. We need more information about males' life histories to understand why natural selection has favored the evolution of polyandrous groups in these monkey species.

MALE-MALE COMPETITION IN NONMONOGAMOUS GROUPS

For males in nonmonogamous groups, reproductive success depends on their ability to gain access to groups of unrelated females and to obtain matings with receptive females.

As we explained in Chapter 6, males often leave their natal groups at puberty and attempt to join new groups. When females are philopatric, male dispersal is obligatory. Dispersal is often a dangerous and stressful time for males, as the reading in this chapter explains. In some species, males disperse alone and spend some time on their own before they join new groups. During this period, males are likely to become more vulnerable to predators and may have trouble gaining access to desirable feeding sites.

Males can reduce the costs of dispersal in several ways. First, before they leave their natal groups, they can scope out neighboring groups when they come close. This may give young males an opportunity to size up the potential prospects (number of available females) and the competition (number and size of adult males). Second, they can transfer to neighboring groups that contain familiar males from their natal groups. In Amboseli, Kenya, vervet males often join neighboring groups that kin or former group members have already joined. Third, males can migrate together with peers or join all-male bands when they leave their natal groups.

In species that normally form one-male groups, males compete actively to establish residence in groups of females.

In primate species that form one-male groups, resident males face persistent pressure from nonresidents. In the highlands of Ethiopia, gelada baboons challenge resident males and attempt to take over their social groups, leading to fierce confrontations that may last for several days (Figure 7.29). Robin Dunbar of Liverpool University has estimated that half of the males involved in aggressive takeover attempts are seriously wounded. Among Hanuman langurs, males form all-male bands that collectively attempt to oust resident males from bisexual (coed) groups. Once they succeed in driving out the resident male, the members of the all-male band compete among themselves for sole access to the group of females. One consequence of this competition is that male tenure in one-male groups is often short.

FIGURE 7.29

Most gelada groups contain only one male. Males sometimes attempt to take over groups and oust the resident male; in other cases males join groups as followers and establish coresidence. Takeovers are risky because they do not always succeed and males are sometimes badly injured.

Dangers of Dispersal

Some primates, such as orangutans, lead solitary lives, meeting only for the occasional mating. But the average group of primates is supremely social, whether it is a family of a dozen gorillas living in a mountain rain forest, a band of twenty langur monkeys on the outskirts of an Indian village, or a troop of one hundred baboons in the African grasslands. In such groups an infant is born into a world filled with relatives, friends, and adversaries, surrounded by intrigue, double-dealing, trysts, and heroics—the staples of great small-town gossip. Pretty heady stuff for the kid that's learning whom it can trust and what the rules are, along with its species' equivalent of table manners.

Most young primates are socialized in this way quite effectively, and home must seem homier all the time. Then, inevitably, a lot of the young must leave the group and set out on their own. It is a simple fact driven by genetics and evolution: if everyone stayed on, matured, and reproduced there, and if their kids stayed on, and their kids' kids too, then ultimately everyone would be pretty closely related. You would have the classic problems of inbreeding—lots of funny-looking kids with six fingers and two tails (as well as more serious genetic problems). Thus essentially all social primates have evolved mechanisms for adolescent emigration from one group to another. Not all adolescents have to leave. The problem of inbreeding is typically solved so long as all the adolescents of one sex go and make their fortune elsewhere; the members of the other sex can remain at home and mate with the newcomers immigrating to the group. In chimpanzees and gorillas, it is typically the females who leave for new groups, while the males stay home with their mothers. But among most Old World monkeys—baboons, macaques, langurs—it is the males who make the transfer. Why one sex transfers in one species but not in another is a complete mystery, the sort that keeps primatologists happily arguing with each other ad nauseam.

This pattern of adolescent transfer solves the specter of inbreeding. To look at it just as a solution to an evolutionary problem is mechanistic, however. You can't lose sight of the fact that those are real animals going through the harrowing process. It is a remarkable thing to observe: every day, in the world of primates, someone young and frightened picks up, leaves Mommy and everyone it knows, and heads off into the unknown.

The baboons that I study each summer in East Africa provide one of the best examples of this pattern of adolescent transfer. Two troops encounter each other at midday at some sort of natural boundary—a river, for example. As is the baboon propensity in such settings, the males of the two troops carry on with a variety of aggressive displays, hooting and hollering with what they no doubt hope is a great air of menace. Eventually everyone gets bored and goes back to eating and lounging, ignoring the interlopers on the other side of the river. Suddenly you spot the kid—some adolescent in your troop. He stands there at the river's edge, absolutely riveted. New baboons, a whole bunch of 'em! He runs five steps toward them, runs four back, searches among the other members of his troop to see why no one else seems mesmerized by the strangers. After endless contemplation, he gingerly crosses the river and sits on the very edge of the other bank, scampering back down in a panic should any new baboon so much as glance at him.

A week later. When the troops run into each other again, the kid repeats the pattern. Except this time he spends the afternoon sitting at the edge of the new troop. At the next encounter he follows them for a short distance before the anxiety becomes too much and he turns back. Finally, one brave night, he stays with them. He may vacillate awhile longer, perhaps even ultimately settling on a third troop, but he has begun his transfer into adulthood.

And what an awful experience it is—a painfully lonely, peripheralized stage of life. There are no freshman orientation weeks, no cohorts of newcomers banding together and covering their nervousness with bravado. There is just a baboon kid, all alone on the edge of a new group, and no one there could care less about him. Actually, that is not true—there are often members of the new troop who pay quite a lot of attention, displaying some of the least-charming behavior seen among social primates, and most reminiscent of that of their human cousins. Suppose you are a low-ranking member of that troop, a puny kid a year or so after your own transfer. You spend most of your time losing fights, being pushed around, having food ripped off from you by someone of higher rank. You have a list of grievances a mile long and there's little you can do about it. Sure, there are youngsters in the troop that you could harass pretty successfully, but if they are pretransfer age, their mothers—and maybe their fathers and their whole extended family—will descend on you like a ton of bricks. Then suddenly, like a gift from heaven, a new, even punier kid shows up: someone to take it out on. (Among chimps, where females do the transferring, the same thing occurs; resident females are brutally aggressive toward the new female living on the group's edge.)

Yet that's only the beginning of a transfer animal's problems. When, as part of my studies of disease patterns among baboons, I anesthetize and examine transfer males, I find that these young animals are teeming with parasites. There is no longer anyone to groom them, to sit with them and methodically clean their fur, half for hygiene, half for friendship. And if no one is interested in grooming a recent transfer animal, certainly no one is interested in anything more intimate than that—it's a time of life where sex is mostly just practicing. The young males suffer all the indignities of being certified primate nerds.

They are also highly vulnerable. If a predator attacks, the transfer animal—who is typically peripheral and exposed to begin with—is not likely to recognize the group's signals and has no one to count on for his defense. I witnessed an incident like this once. The unfortunate animal was so new to my group that he rated only a number, 273, instead of a name. The troop was meandering in the midday heat and descended down the bank of a dry streambed into some bad luck: a half-asleep lioness. Panic ensued, with animals scattering every which way as the lioness stirred—while Male 273 stood bewildered and terribly visible. He was badly mauled and, in a poignant act, crawled for miles to return to his former home troop to die near his mother.

In short, the transfer period is one of the most dangerous and miserable times in a primate's life. Yet, almost inconceivably, life gets better. One day an adolescent female will sit beside our nervous transfer male and briefly groom him. Some afternoon everyone hungrily descends on a tree in fruit and the older adolescent males forget to chase the newcomer away. One morning the adolescent and an adult male exchange greetings (which, among male baboons, consists of yanking on each other's penis, a social gesture predicated on trust if ever I saw one). And someday, inevitably, a terrified new transfer male appears on the scene, and our hero, to his perpetual shame perhaps, indulges in the aggressive pleasures of finding someone lower on the ladder to bully.

In a gradual process of assimilation, the transfer animal makes a friend, finds an ally, mates, and rises in the hierarchy of the troop. It can take years. Which is why Hobbes was such an extraordinary beast.

Three years ago my wife and I spent the summer working with a baboon troop at Amboseli National Park in Kenya, a research site run by the behavioral biologists Jeanne and Stuart Altmann from the University of Chicago. When I'm in the field, I study the relationships between a baboon's rank and personality, how its body responds to stress, and what sorts of stress-related diseases it acquires. In order to obtain the physiological data—blood samples to gauge an animal's stress hormone levels, immune system function, and so on—you have to anesthetize the baboon for a few hours with the aid of a small aluminum blowgun and drug-filled darts. Fill the dart with the right amount of anesthetic, walk up to a baboon, aim, and blow, and he is snoozing five minutes later. Naturally it's not quite that simple—you can only dart someone in the mornings (to control for the effect of circadian rhythms on the animal's hormones). You must ensure there are no predators around to shred the guy, and you must make certain he doesn't climb a tree before passing out. And most of all, you have to dart and remove him from the troop when none of the other baboons are looking so that you don't alarm them and disrupt their habituation to the scientists. So essentially, what I do with my college education is creep around in the bushes after a bunch of baboons, waiting for the instant when they are all looking the other way so that I can zip a dart into someone's tush.

It was about halfway through the season. We were just beginning to know the baboons in our troop (named Hook's troop, for a long-deceased matriarch) and were becoming familiar with their daily routine. Each night they slept in their favorite grove of trees; each morning they rolled out of bed to forage for food in an open savannah strewn with volcanic rocks tossed up aeons ago by nearby Mount Kilimanjaro. It was the dry season, which meant the baboons had to do a bit more walking than usual to find food and water, but there was still plenty of both, and the troop had time to lounge around in the shade during the afternoon heat. Dominating the social hierarchy among the males was an imposing character named Ruto, who had joined the troop a few years before and had risen relatively quickly in the ranks. Number two was a male named Fatso, who had had the misfortune of being a rotund adolescent when he'd transferred into the troop years before and was named by a callous researcher. Fat he was no longer. Now a muscular prime-age male, he was Ruto's most obvious competitor, though still clearly subordinate. It was a fairly peaceful period for the troop; mail was delivered regularly and the trains ran on time.

Then one morning we arrived to find the baboons in complete turmoil. It is not an anthropomorphism to say that everyone was mightily frazzled. There was a new transfer male, and not someone meekly scurrying about the periphery. He was in the middle of the troop, raising hell—threatening and chasing everyone in sight. This nasty, brutish animal was soon named Hobbes (in deference to the seventeenth-century English philosopher who described the life of man as solitary, poor, nasty, brutish, and short).

This is an extremely rare, though not unheard of, event among baboons. The transfer male in such cases is usually a big, muscular, intimidating kid. Maybe he is older than the average seven-year-old transfer male, or perhaps this is his second transfer and he picked up confidence from his first emigration. Maybe he is the cocky son of a high-ranking female in his old troop. In any case, the rare animal with these traits comes on like a locomotive, and he often gets away with it for a time. As long as he keeps pushing, it will be a while before any of the resident males works up the nerve to confront him. Nobody knows him yet; thus no one knows if he is asking for a fatal injury by being the first fool to challenge this aggressive maniac.

This was Hobbes's style. The intimidated males stood around helplessly. Fatso discovered all sorts of errands he had to run elsewhere. Ruto hid behind females. No one else was going to take a stand. Hobbes rose to the number one rank in the troop within a week.

Despite his sudden ascendancy, Hobbes's success wasn't going to last forever. He was still a relatively inexperienced kid, and eventually one of the bigger males was bound to cut him down to size. Hobbes had about a month's free ride. At this point he did something brutally violent but which had a certain grim evolutionary logic. He began to selectively attack pregnant females. He beat and mauled them, causing three out of four to abort within a few days.

One of the great clichés of animal behavior in the context of evolution is that animals act for the good of the species. This idea was discredited in the 1960s but continues to permeate *Wild Kingdom*-like versions of animal behavior. The more accurate view is that animals usually behave in ways that maximize their own reproduction and the reproduction of their close relatives. This helps explain extraordinary acts of altruism and self-sacrifice in some circumstances, and sickening aggression in others. It's in this context that Hobbes's attacks make sense. Were those females to carry through their pregnancies and raise their offspring, they would not be likely to mate again for two years—and who knows where Hobbes would be at that point. Instead he harassed the females into aborting, and they were ovulating again a few weeks later. Although female baboons have a say in whom they mate with, in the case of someone as forceful as Hobbes they have little choice: within weeks Hobbes, still the dominant male in the troop, was mating with two of those three females. (This is not to imply that Hobbes had read his textbooks on evolution, animal behavior, and primate obstetrics and had thought through this strategy, any more than animals think through the logic of when they should reach puberty. The wording here is a convenient shorthand for the more correct way of stating that his pattern of behavior was almost certainly an unconscious, evolved one.)

As it happened, Hobbes's arrival on the scene—just as we had darted and tested about half the animals for our studies—afforded us a rare opportunity. We could compare the physiology of the troop before and after his tumultuous transfer. And in a study published with Jeanne Altmann and Susan Alberts, I documented the not very surprising fact that Hobbes was stressing the bejesus out of these animals. Their blood levels of cortisol (also known as hydrocortisone), one of the hormones most reliably secreted during stress, rose significantly. At the same time, their numbers of white blood cells, or lymphocytes, the sentinel cells of the immune system that defend the body against infections, declined markedly, another highly reliable index of stress. These stress-response markers were most pronounced in the animals receiving the most grief from Hobbes. Unmolested females had three times as many circulating lymphocytes as one poor female who was attacked five times during those first two weeks.

An obvious question: Why doesn't every new transfer male try something as audaciously successful as Hobbes's method? For one thing, most transfer males are too small at the typical transfer age of seven years to intimidate a gazelle, let alone an eighty-pound adult male baboon. (Hobbes, unusually, weighed a good seventy pounds.) Most don't have the personality needed for this sort of unpleasantry. Moreover, it's a risky strategy, as someone like Hobbes stands a good chance of sustaining a crippling injury early in life.

But there was another reason as well, which didn't become apparent until later. One morning, when Hobbes was concentrating on whom to hassle next and paying no attention to us, I managed to put a dart into his haunches. Months later, when examining his blood sample in the laboratory, we found that Hobbes had among the highest levels of cortisol in the troop and extremely low lymphocyte counts, less than one-quarter the troop average. (Ruto and Fatso, conveniently sitting on the sidelines, had three and six times as many lymphocytes as Hobbes had.) The young baboon was experiencing a massive stress response himself, larger even than those of the females he was traumatizing, and certainly larger than is typical of the other, meek transfer males I've studied. In other words, it doesn't come cheap to be a bastard twelve hours a day—a couple of months of this sort of thing is likely to exert a physiological toll.

As a postscript, Hobbes did not hold on to his position. Within five months he was toppled, dropping down to number three in the hierarchy. After three years in the troop, he disappeared into the sunset, transferring out to parts unknown to try his luck in some other troop.

SOURCE: From pp. 78–86 in Robert M. Sapolsky, 1997, *The Trouble with Testosterone*, Scribner, New York. Reprinted by permission of Scribner, a division of Simon & Schuster.

Residence in one-male groups does not always ensure exclusive access to females.

Even though we generally assume that residents of one-male, multifemale groups face little competition over access to females within their groups, this is not always the case. In patas and blue monkeys, for example, researchers have discovered that the resident male is sometimes unable to prevent other males from associating with the group and mating with sexually receptive females. Such incursions are concentrated during the mating season and may last for hours, days, or weeks and involve one or several males.

Some primate species form both one-male and multimale groups, depending on the circumstances. For example, Teresa Pope and Carolyn Crockett have found that in Venezuelan forests that are relatively sparsely populated by red howlers, and where it is relatively easy to establish territories, one-male groups predominate. But when the forests become more densely populated, and dispersal opportunities are more limited, males adopt a different strategy. They pair up with other males and jointly defend access to groups of females. These partnerships enable males to defend larger groups of females and to maintain residence in groups longer (Figure 7.30). Similarly, Hanuman langurs sometimes live in one-male groups and sometimes live in multimale groups.

For males in multimale groups, conflict arises over group membership and access to receptive females.

MALE REPRODUCTIVE TACTICS

In multimale groups, there is more competition over gaining access to mating partners than establishing membership. Nonetheless, it is not necessarily easy to join a new group; no one puts out the welcome mat. in some macaque species, males hover near the periphery of social groups, avoid aggressive challenges by resident males, and attempt to ingratiate themselves with females. In chacma baboons, immigrant males sometimes move directly into the body of the group and engage high-ranking resident males in prolonged vocal duels and chases. Although there may be conflict when males attempt to join nonnatal groups, males spend most of their adult lives in groups containing both males and females.

In multimale groups, males often compete directly over access to receptive females. Sometimes males attempt to drive other males away from females, to interrupt copulations, and to prevent other males from approaching or interacting with females. More often, however, male-male competition is mediated through dominance relationships that reflect male competitive abilities. These relationships are generally established in contests that can involve threats and stereotyped gestures but that can also lead to escalated conflicts in which males chase, wrestle, and bite one another (Figures 7.31 and 7.32). In many cases, the outcomes of encounters between particular pairs of males are relatively predictable from day to day, and males can be ordered in a linear dominance hierarchy. Male fighting ability and dominance rank are generally closely linked to physical condition, as prime-age males in good physical condition are usually able to dominate others. However, the correlation between physical abilities and dominance rank is imperfect, as old males can sometimes dominate younger and stronger males.

It seems logical that male dominance rank would correlate with male reproductive success in multimale groups, but this conclusion has been energetically debated. The issue has been difficult to resolve as it is difficult to infer paternity from behavioral data. It is hard to determine which male fathered a particular infant because females often mate with multiple males when they are receptive, clandestine matings may be overlooked, and observations typically end at sunset whereas sexual activity may extend into the night. However, new genetic techniques make it possible to assess paternity with a much higher degree of precision. As more and more genetic information about paternity has become available, the links between male dominance rank and reproductive success have become stronger. For example, the team of primatologists studying Hanuman langurs at Ramnagar in India recently completed paternity analyses in several multimale groups. The number of adult males in these groups varied from two to eight. If all males in multimale groups were equally likely to sire offspring, then the top-ranking males would have sired about eight infants. However, the top-ranked males sired 17 infants, more than twice as many as expected (Figure 7.33). Genetic analyses of paternity in free-ranging baboons, long-tailed macaques, howler monkeys, and patas monkeys also show a positive relationship between male dominance rank and reproductive success.

FIGURE 7.30

Red howlers live in groups that contain two to four females and one or two mature males. When habitats are saturated and opportunities to form new groups are limited, males pair with other males and jointly defend access to groups of females. These partnerships enable males to defend larger groups of females and to maintain residence in groups for longer periods.

FIGURE 7.31

Male baboons compete over access to an estrous female.

There is growing evidence of substantial variation in the reproductive success of males over the course of their lifetimes.

As we noted earlier, male fitness is expected to vary considerably more than female fitness. Correlations between male rank and reproductive success suggest that male fitness does vary. However, we know that male dominance rank changes over time;

FIGURE 7.32

(a) A male baboon displays his canines as he threatens a rival. (b) Males' canines are formidable weapons, and males sometimes sustain serious injuries in fights with other males. This male baboon was injured in a fight with another male.

(a)　　　　　　　　　　　　　　　　　　　(b)

males may be high ranking and reproductively successful while in their prime, but low ranking and reproductively unsuccessful the rest of their lives. Moreover, male tenure in one-male groups is often quite short, and competition for these positions is intense. If at some point in their lives all males are either high ranking in multimale groups or residents in one-male groups, this might even out any variation in male fitness. However, if there is variation in the likelihood of attaining high rank or maintaining high rank, this will produce real variation in lifetime fitness among males. Thus, we need to follow individual males throughout their lives to evaluate their fitness.

INFANTICIDE

Infanticide is a sexually selected male reproductive strategy.

In the early 1970s, a young graduate student from Harvard University named Sarah Blaffer Hrdy traveled to India to study Hanuman langurs. She was intrigued by what

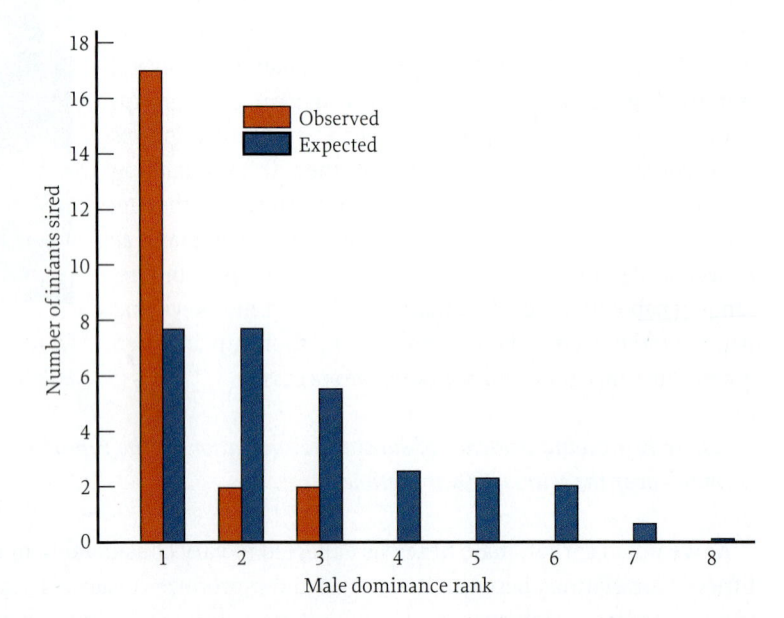

FIGURE 7.33

In Hanuman langur groups at Ramnagar, high-ranking males father the majority of infants in their groups, far more than expected if all males were equally likely to sire infants.

she had read about these animals. Phyllis Jay, who observed Hanuman langurs in central and northern India in 1958–59, described them as peaceful and unaggressive animals. She wrote, "Relations among adult male langurs are relaxed. Dominance is relatively unimportant.... Aggressive threats and fighting are uncommon." This account differed sharply from the reports of Japanese primatologist Yukimaru Sugiyama and his colleagues, who studied Hanuman langurs near Dharwar in southern India a few years later. Sugiyama saw a band of males chase the resident male away from his troop, and then one of the members of the band took control of the troop and drove away all the other males. Shortly after he drove off his rivals, the new resident male attacked and killed the six infants in the group.

Hrdy wondered what accounted for the discrepancy in these accounts (Figure 7.34). Were langurs peaceful creatures whose behavior was pathologically distorted by high levels of crowding near Dharwar? Or were langurs aggressive animals who systematically killed the offspring of their rivals? Was infanticide a widespread occurrence in langur troops or an isolated aberrant incident in the Dharwar troops? To answer these questions, Hrdy began her own study of Hanuman langurs near Mt. Abu in northern India. Over the course of a four-year period, she recorded changes in male membership of several troops and tracked the fate of the infants in these groups.

Hrdy's observations led her to suggest that infanticide is an evolved strategy that enhances male reproductive success, not a pathological response to overcrowding. Hrdy based her hypothesis on the following reasoning. When a female langur gives birth to an infant, she nurses it for a number of months and does not become pregnant again for at least a year. However, after the death of an infant, lactation ends abruptly and females resume cycling. Thus, the death of nursing infants hastens the resumption of maternal receptivity. A male who takes over a group may benefit from killing nursing infants because their deaths cause their mothers to become sexually receptive much sooner than they would otherwise. Since male tenure in langur groups is typically just over two years, and interbirth intervals last nearly three years if infants survive, infanticide may substantially increase a new resident's mating opportunities.

Hrdy's hypothesis, which has become known as the **sexual selection infanticide hypothesis,** was controversial. Some researchers were reluctant to accept the idea that lethal aggression might be adaptive. They insisted that infanticide was pathological and only occurred when langurs lived in disturbed habitats at high density. Some were skeptical of Hrdy's evidence because there were few eyewitness accounts of infanticide in langurs, and she relied heavily on circumstantial evidence that linked male takeovers to infant disappearances. However, Hrdy began to comb the literature to test her ideas. Researchers had reported infant injuries, disappearances, or deaths after changes in male residence in a number of primate species, including gorillas, hamadryas baboons, and howler monkeys. Looking beyond the primates, Hrdy found evidence of similar patterns in rodents and lions.

Intrigued by these patterns, behavioral ecologists began to collect data to test Hrdy's hypothesis. Over the last 25 years, infanticide by males has been reported in approximately 40 primate species. Researchers have now seen at least 60 infanticidal attacks in the wild and have documented many nonlethal attacks on infants by adult males and many more instances in which healthy infants disappeared after takeovers or changes in male rank. Although early studies suggested that sexually selected infanticide was limited to one-male groups, we now know that infanticide also occurs in multimale groups of savanna baboons, langurs, and Japanese macaques.

(a)

(b)

FIGURE 7.34

When Sarah Hrdy began her work, it was not clear whether langurs were (a) aggressive animals who systematically killed the offspring of their rivals or (b) peaceful animals whose behavior was pathologically distorted by high levels of crowding.

This body of data enables researchers to test a number of predictions derived from Hrdy's hypothesis. If infanticide is a male reproductive strategy, then we would expect that 1) infanticide will be associated with changes in male residence or status; 2) males should kill infants whose deaths hasten their mother's resumption of cycling; 3) males should kill other males' infants, not their own; and 4) infanticidal males should achieve reproductive benefits. The data fit all four of these predictions.

Infanticide is linked to changes in male membership and status.

In a recent review of 55 infanticides in free-ranging groups that were actually seen by observers, Carel van Schaik found that 47 infanticides (85%) followed changes in male residence or dominance rank. In one-male groups, infanticide follows takeovers, as Sugiyama and his colleagues originally reported. In multimale groups, infanticide typically follows changes in male residence or dominance rank. For example, in a savanna baboon group in Botswana, observers saw five lethal attacks on infants by males. In three of these cases, the infanticidal male had recently emigrated into the study group and achieved the top-ranking position; twice infants were killed by a resident male shortly after he acquired the top-ranking position (Figure 7.35).

FIGURE 7.35
A male baboon in the Moremi Reserve of the Okavango Delta in Botswana holds the body of an infant that he has just killed. The male had recently acquired the top-ranking position in the group. (Photograph courtesy of Ryne Palombit.)

Infanticide shortens interbirth intervals.

Primate females generally do not conceive while they are lactating, but if their infant dies, they quickly resume cycling. This takes as little as a few days or weeks. The younger an infant is when it dies, the greater the effect of its death on the mother's interbirth interval. This means that we should expect the youngest infants to be at the greatest risk for infanticide. Van Schaik's review of observed infanticides indicates that most involve unweaned infants. Moreover, detailed data from studies of Hanuman langurs and red howlers demonstrate that the youngest infants are at greatest risk after takeovers (Figure 7.36). On average, infanticide reduces the length of interbirth intervals by approximately 25% in langurs and 32% in red howlers. Taken together, these data provide strong evidence that infanticide reduces the duration of the mother's interbirth interval.

Infanticidal males do not kill their own infants.

Infanticide is unlikely to be adaptive if males target their own infants for attack, or even if males are unable to differentiate between their own infants and other males' infants. We do not know whether males can recognize their own infants, but they might be able to avoid harming their own infants if they selectively attacked infants who were conceived before they joined the group or before they rose to high rank. Van Schaik's analysis of 55 observed infanticides shows that 40 infants (73%) were killed by males who were not present in the group at the time the infants were conceived. In 11 (20%) of the remaining cases, the killer was present when the infant was conceived, but was not seen mating in the group. In a detailed long-term study of

FIGURE 7.36

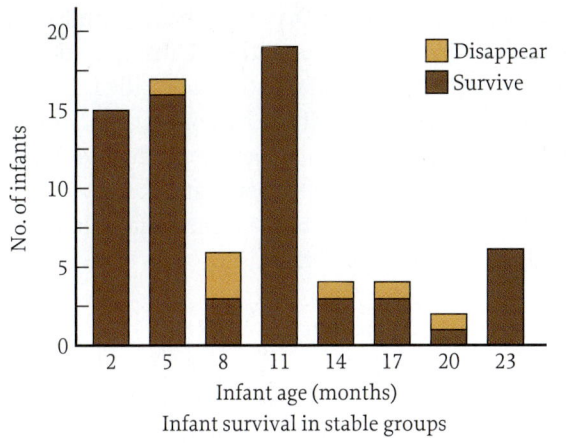

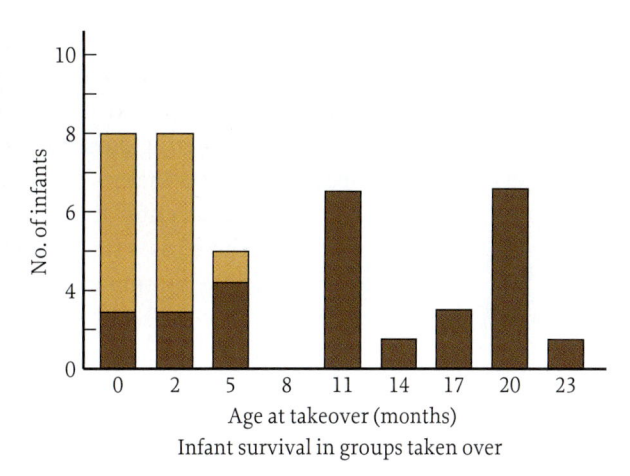

(a) In red howlers in Venezuela, infant mortality is much greater in groups that are taken over by new males than in groups with stable male residents. In groups that are taken over by new males, it is the youngest infants who suffer the highest levels of mortality. (b) An infant, sitting in front of her mother, has been seriously wounded. This infant partially recovered from these wounds but was subsequently attacked by adult males a number of times. Six months later, this infant was the subject of a lethal attack by an adult male. (Photograph courtesy of Carola Borries.)

Hanuman langurs, Volker Sommer documented 55 instances in which infanticide was observed (12 cases) or strongly suspected (43 cases). In 52 (95%) of these cases, a genetic relationship between the killer and his victim was impossible or unlikely.

More compelling evidence comes from two studies in which researchers were able to assess the genetic relationship between males and the infants they attacked. Carola Borries of the State University of New York at Stony Brook and her colleagues documented 24 serious attacks on infants, including one lethal assault, by adult males among the Hanuman langurs near Ramnagar. For 16 of these male-infant pairs, the team obtained DNA samples from both the male and the infant. In all 16 cases, the male was not the infant's father. Joseph Soltis of the National Institutes of Health and his colleagues from the Primate Research Institute in Inuyama, Japan, observed a group of unprovisioned, free-ranging Japanese macaques following a wholesale change in the male dominance hierarchy. They saw one infanticide directly and observed a number of nonlethal attacks on infants by adult males. Altogether, these attacks involved 23 male-infant pairs. In 22 pairs, DNA analyses confirmed that the male was not the father of the infant that he attacked; in the remaining case paternity was not resolved.

Infanticidal males gain reproductive benefits.

The final prediction is that males who commit infanticide should benefit by being able to sire the mother's next infant. Infanticidal males subsequently mated with the mother of the infant that they killed in 25 of the 55 cases that van Schaik compiled. In 13 more cases, males may have mated with the victim's mother. Thus, infanticidal males often gained sexual access to the mothers of the infants that they killed.

Of course, mating does not guarantee paternity. The study by Borries and her colleagues suggests that infanticidal males at Ramnagar do achieve reproductive benefits. They documented five cases in which infants died after being attacked by males; in four of these cases the mother subsequently gave birth to another infant. In all of these cases, the presumed killer sired the mother's next infant.

Infanticide is sometimes a substantial source of mortality.

In some populations, infanticide is known or suspected to be a major source of mortality for infants. Among mountain gorillas in the Karisoke Mountains of Rwanda, savanna baboons in the Moremi Reserve of Botswana, Hanuman langurs in Ramnagar in Nepal, and red howlers in Venezuela, approximately one-third of all infant deaths are due to infanticide.

Females have evolved a battery of responses to infanticidal threats.

While infanticide may enhance male reproductive success, it can have a disastrous effect on females who lose their infants. Thus, we should expect females to evolve counterstrategies to infanticidal threats.

The most obvious counterstrategy would be for females to try to prevent males from harming their infants (Figure 7.37). However, females' efforts to protect their infants are unlikely to be effective. Remember that males are larger than females in nonmonogamous species, and the extent of sexual dimorphism is most pronounced in species that form one-male groups. Thus, in the species in which infanticide is most common, males are much larger than females and have long dangerous canines. Females have little chance of protecting their infants from much larger males who systematically stalk their infants waiting for an opportunity to attack. In fact, mothers are sometimes wounded when their infants are attacked.

Females sometimes enlist males' support against infanticidal males. Ryne Palombit has studied male-female relationships in the Moremi Reserve, where infanticide is a major source of infant mortality. He has found that mothers of young infants form close relationships with adult males. Females groom their male friends at high rates and remain close to them. In return, males are very attentive to their friends' infants and sometimes intervene when these infants are attacked by other males.

Finally, females may try to confuse males about paternity. As we have seen already, males seem to kill infants when there is no ambiguity about their paternity; if females can increase uncertainty about paternity, they may reduce the risk of infanticide. In some primate species, females mate with a number of different males while they are receptive, and in some species, females continue to mate during pregnancy. Both behaviors may reduce certainty about paternity and thereby reduce the risk of infanticide.

If the data are so consistent, why is the idea so controversial?

When Hrdy first proposed the idea that infanticide is an evolved male reproductive strategy, there was plenty of room for skepticism and dispute. The data were limited, and data needed to test the predictions derived from Hrdy's hypothesis had not yet been collected. Now, however, we have observed infanticide in 40 species distributed

FIGURE 7.37
In species like langurs in which infanticide is common, females may lose many of their infants to infanticide. In some species, special relationships with adult males or mating with multiple partners may provide defense against infanticidal attacks.

throughout the primate order, as well as in rodents, birds, and carnivores. We have good evidence that the patterning of infanticidal attacks fits predictions derived from the sexual selection hypothesis; in fact, data from other taxa provide even stronger evidence in support of these predictions than data from primates do. But the controversy has not disappeared. In the last five years a prominent anthropological journal published two papers which contended that there is no support for the hypothesis that infanticide in animal species is a sexually selected male reproductive strategy. Why does the controversy persist?

Volker Sommer, whose own work on infanticide in Hanuman langurs has been attacked by critics of Hrdy's hypothesis, believes that the criticism comes in part from a tendency to commit what is called the "naturalistic fallacy," to assume that what we see in nature is somehow right, just, and inevitable. Thus, critics are concerned that if we accept the idea that infanticide is an adaptive strategy for langurs, baboons, or other primates, it will justify similar behavior in humans. However, as we will discuss more fully in Part Four, the naturalistic fallacy is just that, a fallacy. It is stupid to try to extract moral meaning from the behavior of other animals.

PATERNAL CARE IN NONMONOGAMOUS GROUPS

The extent of male care of offspring varies among nonmonogamous species.

In nonmonogamous species, males generally play little direct role in caring for infants and juveniles. In most cases, they don't carry them, provide food for them, or interact with them often. However, males sometimes contribute to the welfare of immatures in less direct ways.

Males are often quite tolerant of infants and juveniles. Silverbacks sometimes intervene in group conflicts involving infants and juveniles, and generally support the younger of two antagonists when they do so. In brown capuchin groups, the alpha male, who monopolizes matings with receptive females, allows infants to feed near him, thus giving his offspring preferential access to resources (Figure 7.38). In both cases, male tolerance may be a low-cost form of paternal care.

Male baboons sometimes form close relationships ("friendships") with mothers of newborns that are at risk of harassment by high-ranking adults. Males hold, carry, and groom their friends' infants, and sometimes intervene on behalf of immatures when they become involved in agonistic encounters. Ronald Noë found that male baboons are much more likely to intervene on behalf of infants who were conceived after they arrived in the group than infants who were conceived before they arrived. In the multimale groups of langurs near Ramnagar, resident males sometimes defend infants against harassment by other males. Males only protect infants if they were present when the infant was conceived and had copulated with the infants' mother near the time of conception.

FIGURE 7.38

In a group of brown capuchins, the dominant male allows an infant to feed beside him undisturbed. (Photograph by Charles Janson.)

FEMALE MATE CHOICE

Female preferences influence male competitive tactics.

During the last two decades there has been considerable interest in the possibility that female preferences influence mating success. Female mate choice has been demonstrated in a number of animal taxa, including cockroaches, house mice, and frogs. In some cases, females who are able to exercise mate choice have higher fitness than females whose choices are constrained.

FIGURE 7.39

Although male access to females is influenced by the outcome of male-male competition, female choice also plays an important role in some situations. Here a female rhesus monkey evades a male's attempt to copulate with her. (Photograph courtesy of Susan Perry and Joseph Manson.)

Female primates sometimes show preferences, but there is little consensus about the kinds of traits that female primates prefer. For example, female gorillas sometimes leave their groups to join strange males, and their decisions may reflect preferences for less closely related males or for males who can provide more effective protection for their offspring. Free-ranging ring-tailed lemur females apparently prefer to mate with males from neighboring groups and rebuff mating attempts by natal males. Among provisioned groups of rhesus macaques on Cayo Santiago, females prefer unfamiliar, low-ranking males over high-ranking males (Figure 7.39). But in brown capuchin groups, females associate selectively with the highest-ranking male.

Although we don't know the basis for female preferences, there is some evidence that female preferences influence male mating success. In experimental studies of captive hamadryas baboons, Hans Kummer and his colleagues at the University of Zürich have shown that unfamiliar males are unlikely to attempt to disrupt a male-female pair if the female seems closely bonded to her mate. Joseph Soltis and his colleagues have shown that female Japanese macaques clearly prefer some males over others. Female preferences were uncorrelated with male dominance rank, but high-ranking males were still more likely to mate with females than low-ranking ones were. However, genetic analyses revealed that the males who were most attractive to females were most successful in siring their offspring.

Further Reading

Altmann, J. 1980. *Baboon Mothers and Infants.* Harvard University Press, Cambridge, Mass.

Hausfater, G., and S. B. Hrdy, eds. 1984. *Infanticide: Comparative and Evolutionary Perspectives.* Aldine, Hawthorne, N.Y.

Krebs, J. R., and N. B. Davies. 1993. *An Introduction to Behavioural Ecology.* Sinauer Associates, Sunderland, Mass., chaps. 8 and 9.

Nicolson, N. A. 1986. Infants, mothers, and other females. Pp. 330–342 in *Primate Societies*, ed. by B. B. Smuts, D. L. Cheney, R. M. Seyfarth, R. W. Wrangham, and T. T. Struhsaker, University of Chicago Press, Chicago.

Palombit, R. 1999. Infanticide and the evolution of pair bonds in nonhuman primates. *Evolutionary Anthropology* 7:117–129.

van Schaik, C. P., and C. H. Janson, eds. 2000. *Infanticide by Males and Its Implications.* Cambridge University Press, Cambridge.

Study Questions

1. Explain why reproductive success is a critical element of evolution by natural selection. When biologists use the terms *cost* and *benefit,* what currency are they trying to measure?
2. What is the difference between polygyny and polyandry? It seems likely that females might prefer polyandry over polygyny, while males would favor polygyny.

Explain why males and females might prefer different types of mating systems. If this conflict of interest occurs, why is polygyny more common than polyandry?

3. In many primate species, reproduction is highly seasonal. Some researchers have suggested that reproductive seasonality has evolved as a means for females to manipulate their reproductive options. How would reproductive seasonality alter females' options? Why do you think this strategy might be advantageous for females?

4. Imagine that you came upon a species in which males and females were the same size, but males had very large testes in relation to their body size. What would you infer about their social organization? Now suppose you found another species in which males were much larger than females, but males had relatively small testes. What would you deduce about their social system? Why do these relationships hold?

5. Among mammalian species, male fitness is typically more variable than female fitness. Explain why this is often the case. What implications does this have for evolution acting on males and females?

6. What factors influence the reproductive success of females? How do these factors contribute to *variance* in female reproductive success?

7. Biologists use the term *investment* to describe parental care. What elements of the selective forces acting on parental strategies does this term capture?

8. Explain the logic underlying the sexual selection infanticide hypothesis. What predictions follow from this hypothesis? List the predictions and explain why they follow from the hypothesis.

9. In general, infanticide seems to be more common in species that form one-male groups than species that form multimale, multifemale groups or monogamous groups. Explain why this might be the case.

10. In the text, we write, "It is stupid to try to extract moral meaning from the behavior of other animals." Discuss this statement and think about whether you agree or disagree with it.

CHAPTER 8

The Evolution of Social Behavior

KINDS OF SOCIAL INTERACTIONS
ALTRUISM: A CONUNDRUM
KIN SELECTION
   HAMILTON'S RULE
   EVIDENCE OF KIN SELECTION IN PRIMATES
RECIPROCAL ALTRUISM

A social primate lives in a group of known individuals. At one time or another, the other members of the group may become playmates, grooming partners, competitors for food, rivals for mates, allies in aggressive confrontations, caretakers for offspring, collaborators in intergroup encounters, and so on. Even solitary primates, like galagos and orangutans, interact regularly with their neighbors, maturing offspring, and prospective mates.

There is great diversity in the form and frequency of social interactions that occur within and among groups of primates. Just as evolutionary theory provides the framework for understanding the patterning of mating and parenting behaviors in nature, it also provides an essential foundation for understanding the form and distribution of social interactions among individuals within social groups.

Kinds of Social Interactions

Social interactions are behaviors that affect the fitness of more than one individual.

When an animal feeds, travels, or sleeps, it has little impact on the fitness of others. But when two individuals interact, say in competition over a prized resource or in cooperative defense of valuable food items, their activities necessarily involve each other—as collaborators, rivals, or opponents. In these situations, the behavior of one individual directly affects the fitness of the other. Interactions that involve two individuals are called **dyadic** or **pairwise** interactions. Table 8.1 classifies four kinds of pairwise interactions according to their effects on the **actor** (the individual performing the behavior) and the **recipient** (the individual affected by the behavior). An act is said to be beneficial (+) if it increases fitness, and costly or detrimental (−) if it reduces fitness. It is evident that social interactions do not necessarily have the same kinds of effects on actors and on recipients. For example, **selfish** acts are beneficial to the actor and costly to the recipient, while the costs and benefits are reversed for **altruistic** interactions. On the other hand, **mutualistic** interactions are beneficial to both actor and recipient, and **spiteful** interactions are costly to both parties.

Before we go any further, two caveats are in order. First, this classification uses ordinary English words like *altruism* and *spite* because they provide a convenient, easily remembered shorthand for describing the fitness effects of different kinds of social behaviors. However, the technical definitions sometimes differ from the meanings these words have in ordinary usage. For example, an act that is beneficial to the recipient but has no adverse impact on the fitness of the actor might be considered altruistic in common usage but is not altruistic in the biological sense of the word. So, you should be careful to keep the technical definitions in mind when these terms are used.

Second, it is extremely difficult to measure the effects of particular behavioral acts on the fitness of individuals, particularly for long-lived animals like primates. A female may participate in hundreds of grooming sessions over the course of a year and in thousands during her lifetime. At the same time, she has a multitude of different experiences that may influence her reproductive career. As a result, it is virtually impossible to assess the effects of a single behavioral act, or even the effects of a class of behavioral acts, on her reproductive success. Nonetheless, we can make reasonable inferences about the immediate costs and benefits of particular acts based on more general considerations, such as the energy demands or the risks associated with specific behaviors or social interactions. For example, we could assess the caloric value of a food item that a female chimpanzee shares with her daughter or measure the time required to replace the food item with another one of similar value. These kinds of assessments provide a basic estimate of the effects of particular kinds of social interactions on fitness.

TABLE 8.1 A classification of pairwise, or dyadic, social interactions. + indicates a positive effect on fitness, and − a negative impact on fitness.

Case	Actor	Recipient
Selfish	+	−
Mutualistic	+	+
Altruistic	−	+
Spiteful	−	−

Altruism: A Conundrum

Altruistic behavior cannot evolve by ordinary natural selection.

Notice from Table 8.1 that two of the four forms of interactions enhance the fitness of the actor: selfish and mutualistic acts increase the fitness of the actor, and will be favored by natural selection, all other things being equal. Thus, there is no problem in explaining why one female capuchin monkey supplants another from a patch of ripe fruit, or why two chimpanzees jointly corner a red colobus monkey and then share the carcass.

Altruism is a puzzle because it *decreases* the fitness of the individual performing the behavior. According to Darwin's theory, complex adaptations, including behavioral adaptations, must be assembled step by small step, each change favored by natural selection. Thus, while altruistic behaviors might initially arise by accident or as side effects of other behaviors, it seems impossible for complex altruistic behaviors to be assembled by natural selection. Each small genetic change that made it more likely for an individual to perform the behavior would be selected against because of the behavior's negative effects on genetic fitness. The same argument applies to spite. Thus, the existence of either spiteful or altruistic behavior would seem to contradict the fundamental logic of natural selection.

Primates perform altruistic behaviors in nature.

This theoretical conundrum would not be a problem if altruistic interactions were rare or unimportant, and this does seem to be the case for spite. However, there is a lot of evidence that primates, and many other social animals, regularly perform complex altruistic behaviors that play an important role in primate social life. Virtually all social primates groom other group members, picking parasites, scabs, and bits of debris from their hair. When one individual in a group sights a predator, she often gives a distinctive alarm call that alerts the rest of the group. Sometimes two individuals form an alliance against a higher-ranking individual. Occasionally, one individual allows another to share his food. All of these behaviors seem to meet the biological definition of altruism. When it is grooming, the donor expends valuable time and energy that could otherwise be used to look for food, to court prospective mates, or for other vital tasks. The recipient gets a thorough cleaning of parts of her body that she may find difficult to reach and may generally enjoy a period of pleasant relaxation (Figure 8.1). By giving an alarm call, the caller makes herself more conspicuous to predators, while the individuals hearing the warning are able to flee to safety. When animals form alliances, particularly when they come to the defense of other group members, they make themselves vulnerable to retaliation by the ag-

FIGURE 8.1
Gray langurs groom one another. Grooming is usually considered to be altruistic because the groomer expends time and energy when it grooms another animal, and the recipient benefits from having ticks removed from its skin, wounds cleaned, and debris removed from its hair.

gressor while the defended individual may have less chance of being injured or defeated during the encounter (Figure 8.2). Sharing food may reduce the amount of food that the owner eats while increasing the amount of food that the recipient obtains. Thus, in each of these cases, it seems likely that the actor suffers a decrease in fitness while the recipient incurs an increase in fitness, which satisfies the criteria for altruism.

Altruistic behaviors cannot be favored by selection just because they are beneficial to the group as a whole.

You might think that if the average effect of an act on all members of the group is positive, then it would be beneficial for all individuals to perform it. For example, suppose that when one monkey gives an alarm call, the other members of the group benefit, and the total benefits to all group members exceed the cost of giving the call. Then, if every individual gives the call when a predator is sighted, all members of the group will be better off than if no warning calls were ever given.

The problem with this line of reasoning is that it confuses the effect on the group with the effect on the actor. In most circumstances, the fact that alarm calls are beneficial to those hearing them doesn't affect whether the trait alarm calling evolves; all that matters is the effect of giving the alarm call on the caller. To see why, imagine a hypothetical monkey species in which some individuals give alarm calls when they are the first to spot a predator. Monkeys who are warned by the call have a chance to flee. Suppose that one-fourth of the population gives the call when they spot predators ("callers") and three-fourths of the individuals do not give an alarm call in the same circumstance ("noncallers"). (These proportions are arbitrary; we chose them because it's easier to follow the reasoning in examples with concrete numbers.) Let's suppose that in this species the tendency to give alarm calls is genetically inherited.

FIGURE 8.2

Two capuchin monkeys form a coalition against an opponent who is not visible in the photograph. Coalitions are altruistic because the ally puts itself at risk and expends energy when it becomes involved in an ongoing dispute. The animal that receives support may benefit if the fight ends sooner or with less costly consequences. (Photograph by Susan Perry.)

Now we compare the fitness of callers and noncallers. Since everyone in the group can hear the alarm calls and take appropriate action, alarm calls benefit everyone in the group to the same extent (Figure 8.3). Calling has no effect on the *relative* fitness of callers and noncallers because, on average, one-fourth of the beneficiaries will be callers and three-fourths of the beneficiaries will be noncallers, the same proportions we find in the population as whole. Calling reduces the risk of mortality for everyone who hears the call, but it does not change the frequency of callers and noncallers in the population because everyone gains the same benefits. However, callers are conspicuous when they call, which makes them more vulnerable to predators. While all individuals benefit from hearing alarm calls, callers are the only ones who suffer the costs from calling. This means that, on average, noncallers will have a higher fitness than callers. Thus, genes that cause alarm calling will not be favored by selection even if the cost of giving alarm calls is small and the benefit to the rest of the group is large. Instead, selection will favor genes that suppress alarm calling because noncallers have higher fitness than callers. This simple example can be reformulated for grooming, food sharing, coalition formation, and so on, and in each case the conclusion is

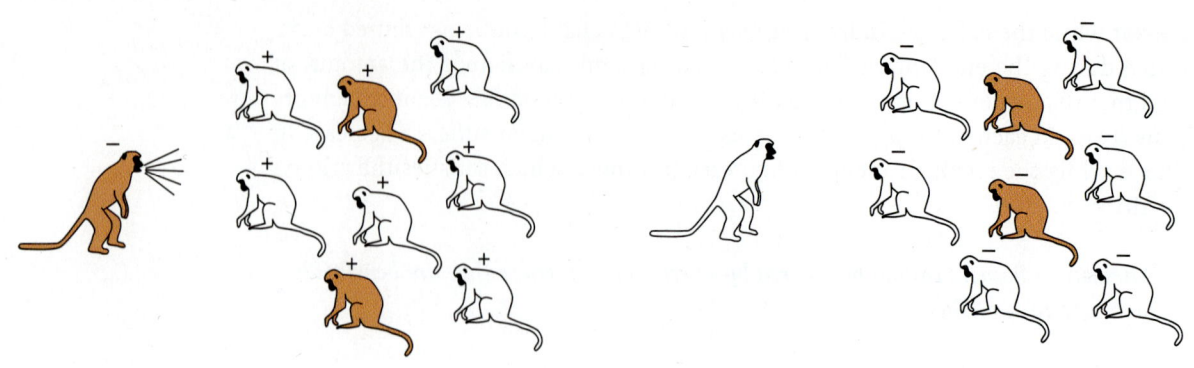

(a) Altruist gives alarm call to group.

(b) Nonaltruist doesn't give alarm call to group.

FIGURE 8.3

Two groups of monkeys are approached by a predator. (a) In one group, there is an individual (pink) who has a gene that makes her call in this context. Giving the call lowers the caller's fitness but increases the fitness of every other individual in the group. Like the rest of the population, one out of four of these beneficiaries also carries the genes for calling. (b) In the second group, the female who detects the predator does not carry the gene for calling and remains silent. This lowers the fitness of all members of the group a certain amount because they are more likely to be caught unaware by the predator. Once again, one out of four is a caller. Although members of the caller's group are better off on average than members of the noncaller's group, the gene for calling is not favored. This is because callers and noncallers in the caller's group both benefit from the caller's behavior, but callers incur some costs. Callers are at a disadvantage relative to noncallers. Thus, calling is not favored, even though the group as a whole benefits.

the same—individual selection is not expected to favor the evolution of these kinds of behaviors (Box 8.1).

Kin Selection

Natural selection can favor altruistic behavior if altruistic individuals are more likely to interact with each other than chance alone would dictate.

If altruistic behaviors can't evolve by ordinary natural selection or by group selection, then how do they evolve? A clear answer to this question did not come until 1964, when the young biologist W. D. Hamilton published a landmark paper. This paper was the first of a series of fundamental contributions Hamilton made to our understanding of the evolution of behavior. There are several different ways to conceptualize Hamilton's basic idea, and we will adopt the approach presented in his original paper.

The argument made in the previous section contains a hidden assumption—that altruists and nonaltruists are equally likely to interact with one another. Thus, we supposed that callers give alarm calls when they hear a predator, no matter who is nearby. This is why callers and noncallers were equally likely to benefit from hearing an alarm call. Hamilton's insight was to see that any process that causes altruists to be more likely to interact with other altruists than they would by chance could facilitate the evolution of altruism.

To see why this is such an important insight, let's modify the previous example by assuming that our hypothetical species lives in groups composed of full siblings—offspring of the same mother and father. The frequency of the calling and noncalling

Box 8.1
Group Selection

Group selection was once thought to be the mechanism for the evolution of altruistic interactions. In the early 1960s a British ornithologist, V. C. Wynne-Edwards, contended that altruistic behaviors like those we have been considering here evolved because they enhanced the survival of whole groups of organisms. Thus, individuals gave alarm calls, despite the costs of becoming more conspicuous to predators, because calling protected the group as a whole from attacks. Wynne-Edwards reasoned that groups that contained more altruistic individuals would be more likely to survive and prosper than groups that contained fewer altruists, and the frequency of the genes leading to altruism would increase.

Wynne-Edwards's argument is logical, because Darwin's postulates logically apply to groups as well as individuals. However, group selection is not an important force in nature because there is generally not enough genetic variation among groups for selection to act on. Group selection can occur if groups vary in their ability to survive and to reproduce, and that variation is heritable. Then group selection may increase the frequency of genes that increase group survival and reproductive success. The strength of selection among groups depends on the amount of genetic variation among groups, just as the strength of selection among individuals depends on the amount of genetic variation among individuals. However, when individual selection and group selection are opposed, and group selection favors altruistic behavior while individual selection favors selfish alternatives, individual selection has a tremendous advantage. This is because the amount of variation among groups is much smaller than the amount of variation among individuals, unless groups are very small and there is very little migration among them. Thus, individual selection favoring selfish behavior will generally prevail over group selection, making group selection an unlikely source of altruism in nature.

genes doesn't change, but their distribution will be affected by the fact that siblings live together (Figure 8.4). If an individual is a caller, then by the rules of Mendelian genetics there is a 50% chance that the individual's siblings will share the genes that cause calling behavior. This means that the frequency of the genes for calling will be higher in groups that contain callers than in the population as a whole, and therefore, more than one-fourth of the beneficiaries will be callers themselves. When a caller gives an alarm call, the audience will contain a higher fraction of callers than the population at large. Thus, the caller raises the average fitness of callers relative to noncallers. Similarly, because the siblings of noncallers are more likely to be noncallers than chance alone would dictate, callers are less likely to be present in such groups than in the population at large. Therefore, the absence of a warning call lowers the relative fitness of noncallers relative to callers.

When individuals interact selectively with relatives, callers are more likely to benefit than noncallers, and the benefits of calling will, all other things being equal, favor the genes for calling. However, we must remember that calling is costly, and this will tend to reduce the fitness of callers. Calling will only be favored by natural selection if the benefits of calling are sufficiently greater than the costs. The exact nature of this tradeoff is specified by what we call Hamilton's rule.

(a) Altruist gives alarm call to siblings.

(b) Nonaltruist doesn't give alarm call to siblings.

FIGURE 8.4
Two groups of monkeys are approached by a predator. Each group is composed of nine sisters. (a) In one group there is a caller (pink), an individual who has a gene that makes her call in this context. Her call lowers her own fitness but increases the fitness of her sisters. (b) In the second group, the female who detects the predator is not a caller and does not call when she spots the predator. As in Figure 8.3, calling benefits the other group members but imposes costs on the caller. However, there is an important difference between the situations portrayed here and in Figure 8.3. Here the groups are made up of sisters, so five of the eight recipients of the call also carry the calling gene. In any pair of siblings, half of the genes are identical because the siblings inherited the same gene from one of their parents. Thus, on average, half of the caller's siblings also carry the calling allele. The remaining four siblings carry genes inherited from the other parent, and like the population as a whole, one out of four of them is a caller. The same reasoning shows that in the group with the noncaller, there is only one caller among the beneficiaries of the call. Half are identical to their sister because they inherited the same noncalling gene from one of their parents; one of the remaining four is a caller. In this situation, callers are more likely to benefit from calling than noncallers, and so calling alters the relative fitness of callers and noncallers. Whether the calling behavior actually evolves depends on whether these benefits are big enough to compensate for the reduction in the caller's fitness.

Hamilton's Rule

Hamilton's theory of kin selection predicts that altruistic behaviors will be favored by selection if the costs of performing the behavior are less than the benefits discounted by the coefficient of relatedness between actor and recipient.

Hamilton's theory of **kin selection** is based on the idea that selection could favor altruistic alleles if animals interact selectively with their genetic relatives. Hamilton's theory also specifies the quantity and distribution of help among individuals. According to **Hamilton's rule,** an act will be favored by selection if

$$rb > c \qquad (1)$$

where

r = the average coefficient of relatedness between the actor and the recipients
b = the sum of the fitness benefits to all individuals affected by the behavior
c = the fitness cost to the individual performing the behavior

The **coefficient of relatedness, r**, measures the genetic relationship between interacting individuals. More precisely, r is the average probability that two individuals acquire the same allele through descent from a common ancestor. Figure 8.5 shows

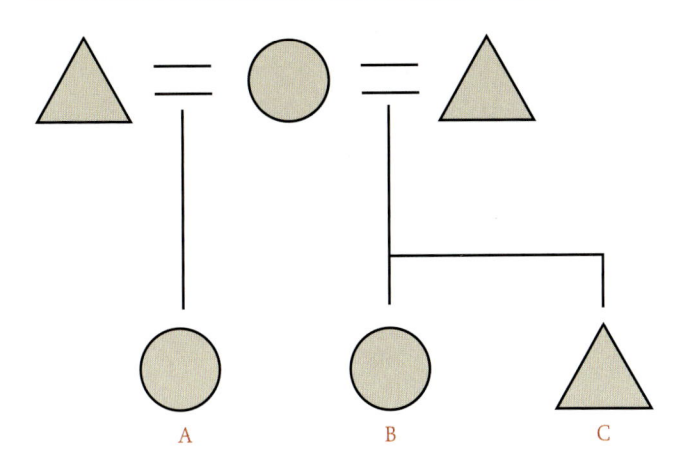

KIN SELECTION

FIGURE 8.5
This genealogy shows how the value of r is computed. Triangles represent males, circles represent females, and the equals sign represents mating. The relationships between individuals labeled in the genealogy are described in the text.

how these probabilities are derived in a simple genealogy. Female A obtains one allele at a given locus from her mother and one from her father. Her half-sister, female B, also obtains one allele at the same locus from each of her parents. The probability that both females obtain the same allele from their mother is obtained by multiplying the probability that female A obtains the allele (0.5) by the probability that female B obtains the same allele (0.5), and equals 0.25. Thus, half-sisters have, on average, a 25% chance of obtaining the same allele from their mothers. Now consider the relatedness between female B and her brother, male C. In this case, note that female B and male C are full siblings: they have the same mother and the same father. The probability of each acquiring the same allele from their mother is still 0.25, but female B and male C might also share an allele from their father. The probability of this event is also 0.25. Thus, the probability that female B and male C share an allele is equal to the sum of 0.25 and 0.25, or 0.5. This basic reasoning can be extended to calculate the degrees of relatedness among various categories of kin (Table 8.2).

Hamilton's rule leads to two important insights: 1) altruism is limited to kin, and 2) closer kinship facilitates more costly altruism.

If you reflect on Hamilton's rule for a while, you will see that it produces two fundamental predictions about the conditions that favor the evolution of altruistic behaviors. First, altruism is not expected to be directed toward nonkin because the

TABLE 8.2 The value of r for selected categories of relatives.

Relationship	r
Parent and offspring	.5
Full siblings	.5
Half-siblings	.25
Grandparent and grandchild	.25
Aunt-uncle and niece-nephew	.25
First cousins	.125
Unrelated individuals	0

FIGURE 8.6

As the degree of relatedness (r) between two individuals declines, the value of the ratio of benefits to costs (b/c) that is required to satisfy Hamilton's rule for the evolution of altruism rises rapidly.

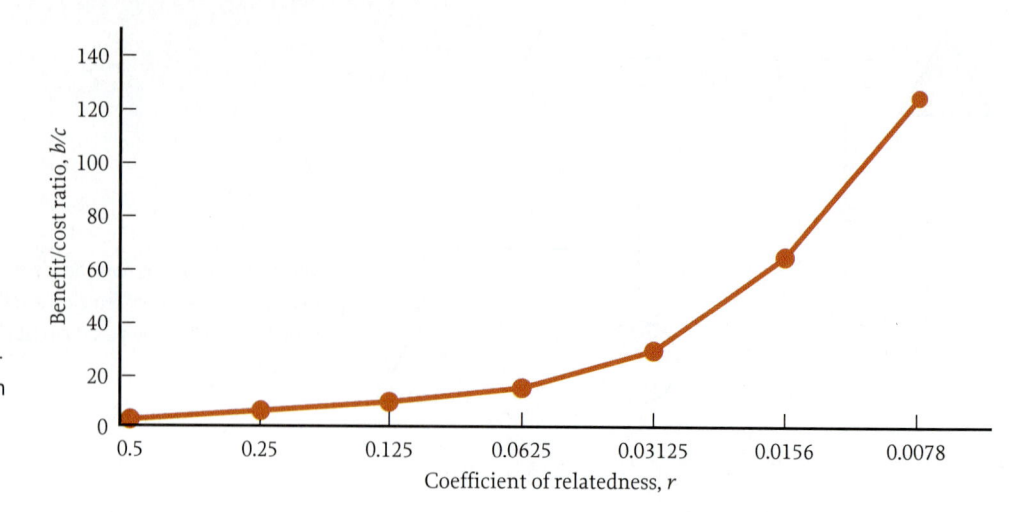

coefficient of relatedness, r, between nonkin is 0. The condition for the evolution of altruistic traits will only be satisfied for interactions between kin, when $r > 0$. Thus, altruists are expected to be nepotistic, showing favoritism toward kin.

Second, close kinship is expected to facilitate altruism. If an act is particularly costly, it is most likely to be restricted to close kin. Figure 8.6 shows how the benefit/cost ratio scales with the degree of relatedness among individuals. Compare what happens when $r = 1/16$ (or 0.0625) and when $r = 1/2$ (or 0.5). When $r = 1/16$, the benefits must be more than 16 times as great as the costs for Hamilton's inequality (Equation 1) to be satisfied. When $r = 1/2$, the benefit needs to be just over twice as large as the costs. All other things being equal, altruism will be more common among close relatives than among distant ones.

It is important to understand that the fractions in Table 8.2 are not equivalent to the quantities in a recipe. If a recipe calls for 2 cups of flour and ¾ cup of sugar, the dutiful cook measures out these amounts. But monkeys don't necessarily apportion altruism in precise amounts according to the coefficients of relatedness. Reality is, as always, more complicated. There are sometimes dramatic asymmetries in the benefit/cost ratios within pairs of relatives which influence the distribution of altruism. For example, in many primate species, it may be much less costly for a mother to defend her infant than vice versa. The inability to recognize certain categories of kin, particularly distant kin or paternal kin, may also limit the distribution of altruism. Finally, it is important to remember that close kinship does not always prevent violence and aggression. In some species of birds, nestlings kill their own siblings to prevent them from competing for resources that parents bring back to the nest. Juvenile primates sometimes jostle with their newborn siblings for their mother's time and attention.

EVIDENCE OF KIN SELECTION IN PRIMATES

A considerable body of evidence suggests that the patterns of many forms of altruistic interactions among primates are largely consistent with predictions derived from Hamilton's rule. Later in this section we consider three examples: food sharing, grooming, and coalition formation. In each case, the behaviors are usually biased toward kin. First, however, we discuss how primates identify relatives.

Primates use contextual cues to recognize maternal relatives.

KIN SELECTION

FIGURE 8.7
(a) Even mothers must learn to recognize their own infants. Here, a female bonnet macaque peers into her infant's face. (b) Primates apparently learn who their relatives are by observing patterns of association among group members. Here, a female inspects another female's infant. (a, Photograph courtesy of Kathy West.)

In order for kin selection to provide an effective mechanism for the evolution of cooperative behavior, animals must be able to distinguish relatives from nonrelatives and close relatives from distant ones. Some organisms are able to recognize their kin by their smell or likeness to themselves. This is called **phenotypic matching.** Others learn to recognize relatives using contextual cues—they use cues such as familiarity and proximity that predict kinship (Figure 8.7). We have generally assumed that primates rely on contextual cues to identify their relatives, but new data suggest that phenotypic matching may also play a role in primate kin recognition.

Mothers seem to make use of contextual cues to recognize their own infants. After they give birth, females repeatedly sniff and inspect their newborns. Nevertheless, if a strange infant is substituted for the female's own infant shortly after birth, most mothers accept the strange infant without any clear sign of having noticed the substitution. But sometime within the first few weeks of life, mothers learn to recognize their infants. After this, females of most species nurse only their own infants and respond selectively to their own infants' distress calls. Although it might seem advantageous for primates to have an innate means of recognizing their young from the moment of birth, there is actually little need for this since young infants spend virtually all of their time in physical contact with their mothers. Thus, mothers are unlikely to confuse their own newborn with another.

Monkeys and apes probably learn to recognize other maternal kin through contact with their mothers. Offspring continue to spend considerable amounts of time with their mothers even after their younger siblings are born. Thus, they have many opportunities to watch their mothers interact with their new brothers and sisters. Similarly, the newborn infant's most common companions are its mother and siblings (Figure 8.8). Because adult females continue to associate with their mothers, infants also become familiar with their grandmothers, aunts, and cousins.

Contextual cues are more useful for recognizing maternal kin than paternal kin.

Close associations between males and females are uncommon in many primate species, and this limits the usefulness of contextual cues for identifying paternal kin. Moreover, there may be considerable uncertainty about paternity in most primate

FIGURE 8.8
Monkey and ape infants grow up surrounded by various relatives. (a) These adult baboon females are mother and daughter. Both have young infants. (b) An adolescent female bonnet macaque carries her younger brother while her mother is recovering from a serious illness.

species. Even in pair-bonded species, like gibbons and callicebus monkeys, females sometimes mate with males from outside their groups. An inability to identify paternal kin would be costly because it would prevent individuals from directing beneficial acts toward certain classes of kin.

A number of researchers have examined the possibility that primates can recognize paternal kin based on phenotypic cues alone. In the laboratory, where matings can be controlled and social groups can be manipulated, researchers are able to examine the effects of familiarity and kinship independently. Several studies produced the same result: monkeys show clear preferences for familiar individuals over unfamiliar individuals, but they don't prefer unfamiliar kin over unfamiliar nonkin. Thus, most primatologists became convinced that primates rely on familiarity, not phenotypic cues, to distinguish kin from nonkin, and therefore cannot recognize paternal kin.

However, new evidence from field studies suggests that the conventional wisdom may be wrong. Jeanne Altmann of Princeton University has pointed out that age may provide a good proxy measure of paternal kinship in species in which a single male typically dominates mating activity within the group. When this happens, all infants born at about the same time are likely to have the same father. Thus, age-mates are likely to be paternal half-siblings. Recent studies suggest that Altmann's logic is correct and that monkeys do use age to identify paternal kin. Anja Widdig of the Humboldt-Universität in Berlin and her colleagues studied kin recognition among rhesus macaques on Cayo Santiago. Females showed strong affinities for maternal half-sisters, as expected. However, females also showed strong affinities for their paternal half-sisters, spending more time grooming and in close proximity to them than to unrelated females. Their affinities for paternal kin seem to be partly based on strong preferences for interacting with age-mates. Thus, monkeys might be able to identify paternal kin by selectively associating with animals of their own age. These data suggest, however, that monkeys may also use phenotypic cues to recognize paternal kin. Thus, females also distinguished *among* their age-mates, preferring paternal half-sisters over unrelated females of the same age.

Food sharing occurs mainly among kin.

In a small number of primate species, adults voluntarily share food items with other group members. **Food sharing,** the unforced transfer of food from one individual to another, generally involves close relatives (Figure 8.9). At Gombe Stream National Park in Tanzania, researchers provided chimpanzees with bananas, a food the chimpanzees coveted but that doesn't grow in their home range (Figure 8.10). William McGrew, now at the University of Miami in Ohio, found that 86% of such exchanges involved maternal kin even though these individuals constituted only 5% of the possible pairs of individuals in the group. Most exchanges involved mothers sharing bananas with their own offspring. However, not all food sharing between chimpanzees involves kin. When male chimpanzees make a kill, they often share meat with other males and with unrelated adult females.

FIGURE 8.9
A female baboon allows her infant to share gum that she has extracted from an acacia tree.

KIN SELECTION

Food sharing also occurs in tamarins and marmosets. These tiny monkeys are omnivores, feeding on fruit, gum and sap, nectar, insects, and small vertebrates. Young marmosets and tamarins have trouble catching large and mobile insect prey and manipulating large fruits, and so older group members selectively share these foods with infants. In some cases, food items are spontaneously offered to infants and juveniles, and in other cases infants use specialized begging vocalizations to solicit food from others. Since tamarins and marmosets live in small monogamous or polyandrous groups, infants are probably closely related to most other group members.

Grooming is also more common among kin than nonkin.

Social **grooming** plays an important role in the lives of most gregarious primates (Figure 8.11). Grooming is likely to be beneficial to the participants in at least two ways. First, grooming serves hygienic functions as bits of dead skin, debris, and parasites are removed and wounds are kept clean and open. Second, grooming may provide a means for individuals to establish relaxed, **affiliative** (friendly) contact and to reinforce social relationships with other group members (Box 8.2). Grooming is costly because the actor expends both time and energy in performing these services. Moreover, Marina Cords of Columbia University has shown that blue monkeys are less vigilant when they are grooming, perhaps exposing themselves to some risk of being captured by predators.

FIGURE 8.10

A female chimpanzee allows her infant to share bananas that she has received at the feeding station in the Gombe Stream Research Center. A veteran Tanzanian field assistant, Hilali Matama, monitors the situation.

(a)

(b)

(c)

(d)

FIGURE 8.11

Some of the many species of primates that groom are (a) capuchin monkeys, (b) blue monkeys, (c) baboons, and (d) gorillas. (Photographs courtesy of: a, Susan Perry; b, Marina Cords; d, John Mitani.)

Box 8.2
How Relationships Are Maintained

Conflict and competition are fundamental features of social life for many primates—females launch unprovoked attacks on unsuspecting victims, males battle over access to receptive females, subordinates are supplanted from choice feeding sites, and dominance relationships are clearly defined and frequently reinforced. Although violence and aggression are not prevalent in all primates (muriquis, for example, are so peaceful that dominance hierarchies cannot be detected), some primates can be charitably characterized as contentious. This raises an intriguing question: how is social life sustained in the face of such relentless conflict? After all, it seems inevitable that aggression and conflict would drive animals apart, disrupt social bonds, and reduce the cohesiveness of social groups. Recently, primatologists have begun to consider this issue.

Social relationships matter to primates. They spend a considerable portion of every day grooming other group members. Grooming is typically focused on a relatively small number of partners and is often reciprocated. Robin Dunbar of Liverpool University contends that in Old World monkeys, grooming has transcended its original hygienic function and now serves as a means to cultivate and maintain social bonds. Social bonds may have real adaptive value to individuals. For example, grooming is sometimes exchanged for support in coalitions, and grooming partners may be allowed to share access to scarce resources.

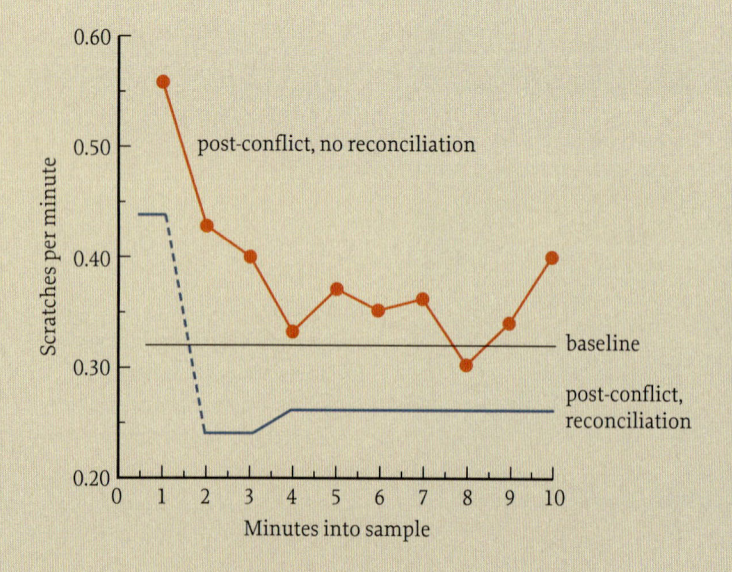

When tensions do erupt into violence, certain behavioral mechanisms may reduce the disruptive effects of conflict on social relationships. After conflicts end, victims often flee from their attackers, an understandable response. However, in some cases, former opponents make peaceful contact in the minutes that follow conflicts. For example, chimpanzees sometimes kiss their former opponents; female baboons grunt quietly to their former victims; and golden monkeys may embrace or groom their former adversaries. The swift transformation from aggression to affiliation prompted Frans de Waal of Emory University to suggest that these peaceful post-conflict interactions are a form of reconciliation, a way to mend relationships that were damaged by conflict. Inspired by de Waal's work, a number of researchers have documented reconciliatory behavior in a wide range of primate species.

Peaceful post-conflict interactions seem to have a calming effect on former opponents. When monkeys are nervous and anxious, rates of certain self-directed behaviors, such as scratching, increase. Thus, self-directed behaviors are a good behavioral index of stress. Felipo Aureli of Liverpool John Moores University and his colleagues at the University of Utrecht and Emory

FIGURE 8.12

Rates of scratching, an observable index of stress, by victims of aggression are elevated over normal levels in the minutes that follow aggressive encounters. However, if some form of affiliative contact (reconciliation) between former opponents occurs during the post-conflict period, rates of scratching drop rapidly below baseline levels. If there is no reconciliatory contact during the first few minutes of the post-conflict period, rates of scratching remain elevated above baseline levels for several minutes. These data suggest that affiliative contact between former opponents has a calming effect. Similar effects on aggressors have also been detected.

University have studied the effects of fighting and reconciliation on the rate of self-directed behaviors. They have found that levels of self-directed behavior, and presumably stress, rise above baseline levels after conflicts. Both victims and aggressors seem to feel the stressful effects of conflicts. If former opponents interact peacefully in the minutes that follow conflicts, rates of self-directed behavior fall rapidly to baseline levels (Figure 8.12). If they do not reconcile, rates of self-directed behavior remain elevated above baseline levels for several minutes longer. If reconciliation provides a means to preserve social bonds, then we would expect primates to reconcile selectively with their closest associates. In a number of groups, former opponents who have strong social bonds are most likely to reconcile. Kin also reconcile at high rates in some groups, even though some researchers have argued that kin have little need to reconcile because their relationships are unlikely to be frayed by conflict.

Reconciliation may also play a role in resolving conflicts among individuals who do not have strong social bonds. Like many other primates, female baboons are strongly attracted to newborn infants and make persistent efforts to handle them. Mothers reluctantly tolerate infant handling, but do not welcome the attention. Female baboons reconcile at particularly high rates with the mothers of young infants, even when they do not have close relationships with them. Reconciliation greatly enhances the likelihood that aggressors will be able to handle their former victims' infants in the minutes that follow conflicts. Thus, in this case, reconciliation seems to be a means to an immediate end but not a means to preserve long-term relationships.

Grooming is more common among kin, particularly mothers and their offspring, than among nonkin. For example, Ellen Kapsalis and Carol Berman of the State University of New York at Buffalo recently documented the effect of maternal relatedness among rhesus macaques on Cayo Santiago Island. in this population, females groom close kin at higher rates than nonkin, and close kin were groomed more often than distant kin were (Figure 8.13). It is interesting to note that as relatedness declined, the differences in the proportions of time spent grooming kin and nonkin were essentially eliminated. This may mean that monkeys cannot recognize more distant kin or that the conditions of Hamilton's rule ($rb > c$) are rarely satisfied for distant kin.

Coalition formation occurs most frequently among close kin.

Most disputes in primate groups involve two individuals. Sometimes, however, several individuals jointly attack another individual, or one individual comes to the support of another individual involved in an ongoing dispute (Figure 8.14). We call these events **coalitions** or **alliances.** Support is likely to be beneficial to the individual that receives aid because it alters the balance of power among the original contestants.

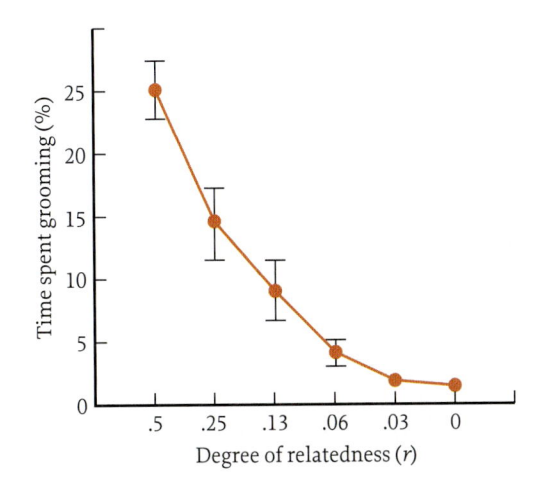

FIGURE 8.13

Rhesus monkeys on Cayo Santiago Island groom close relatives more often than they groom distant relatives or nonkin.

FIGURE 8.14
Two baboons form an alliance against an adult female.

The beneficiary may be more likely to win the contest or less likely to be injured in the confrontation. At the same time, however, intervention may be costly to the supporter who expends time and energy and risks defeat or injury when it becomes involved. Hamilton's rule predicts that support will be preferentially directed toward kin and that the greatest costs will be expended on behalf of close relatives.

Many studies have shown that support is selectively directed toward close kin. Female pigtail macaques defend their offspring and close kin more often than they defend distant relatives or unrelated individuals (Table 8.3). Females run some risk when they participate in coalitions, particularly when they are allied against higher-ranking individuals. Coalitions against high-ranking individuals are more likely to result in retaliatory attacks against the supporter than are coalitions against lower-ranking individuals. Female macaques are much more likely to intervene against higher-ranking females on behalf of their own offspring than on behalf of unrelated females or juveniles. Thus, macaque females take the greatest risks on behalf of their closest kin.

Kin-based support in conflicts has far-reaching effects on the social structure of many primate groups. We have evidence of this from studying macaque, vervet, and baboon groups.

Maternal support in macaques, vervets, and baboons influences the outcome of aggressive interactions and dominance contests. Initially, an immature monkey is able to defeat older and larger juveniles only when its mother is nearby. Eventually, regardless of their age or size, juveniles are able to defeat everyone their mothers can defeat, even when the mother is some distance away. Maternal support contributes directly to several remarkable properties of dominance hierarchies within these species:

- Maternal rank is transferred with great fidelity to offspring, particularly daughters. For example, in a group of baboons at Gilgil, Kenya, maternal rank is an almost perfect predictor of the daughter's rank (Figure 8.15).

TABLE 8.3 Captive pigtail macaques support close kin more frequently than distant kin. Individuals related by $r = 0.5$ include parents and offspring and full siblings. Pairs of individuals related by $r = 0.25$ include half-siblings with each other, and grandparents and grandoffspring. Pairs of individuals related by $r = 0.125$ include cousins. (From Table 1 in A. Massey, 1977, Agonistic aids and kinship in a group of pigtail macaques, *Behavioral Ecology and Sociobiology* 2:31–40.)

	DEGREES OF RELATEDNESS		
	0.5	0.25	0.125
Number of individuals	38	156	48
Number of aids	173	164	13
Number of aids per pair	4.55	1.05	0.27
Number of aids per aggressive encounter	0.15	0.04	0.01

KIN SELECTION

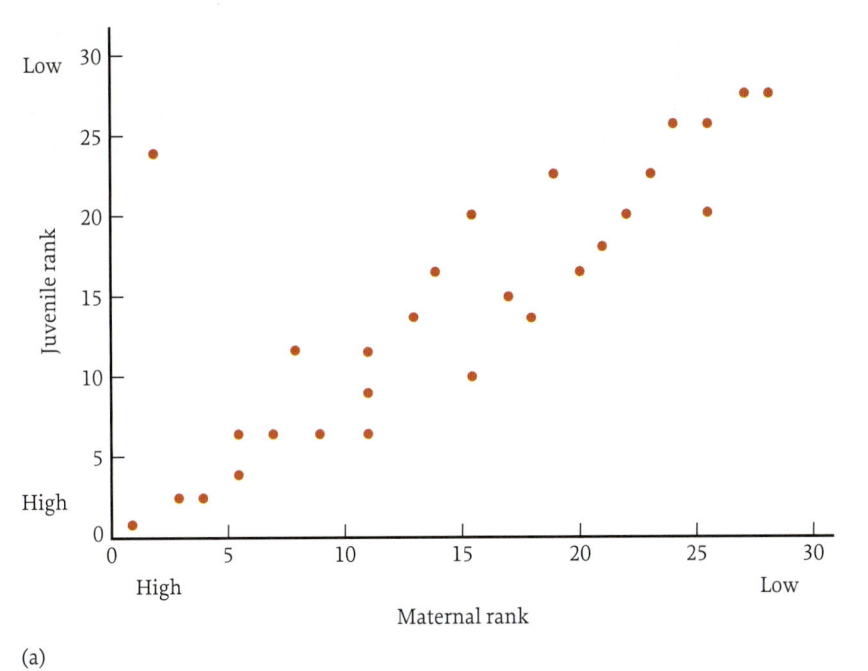

FIGURE 8.15

Juvenile female baboons acquire ranks very similar to their mother's rank. (a) The anomalous point at maternal rank 2 belongs to a female whose mother died when she was an infant. (b) Here, the dominant female of Alto's group is flanked by her two daughters, ranked 2 and 3.

- Maternal kin occupy adjacent ranks in the dominance hierarchy, and all the members of one **matrilineage** (maternal kin group) rank above or below all members of other matrilineages.
- Ranking within matrilineages is often quite predictable. In most cases, mothers outrank their daughters, and younger sisters outrank their older sisters.
- Female dominance relationships are amazingly stable over time. In many groups, they remain the same over months and sometimes over years. The stability of dominance relationships among females may be a result of the tendency to form alliances in support of kin.

Kin-biased support also plays an important role in the reproductive strategies of red howler males.

Behavioral and genetic studies of red howlers in Venezuela conducted by Teresa Pope have shown that kinship influences males' behavior in important ways. Red howlers live in groups that contain two to four females and one or two males. Males sometimes join up with migrant, extragroup females and help them establish new territories. Once such groups are established, resident males must defend their position and their progeny from infanticidal attacks by alien males. When habitats are crowded, males can only gain access to breeding females by taking over established groups and evicting male residents. This is a risky endeavor, as males are often injured in takeover attempts. Moreover, as habitats become more saturated and dispersal opportunities become more limited, males tend to remain in their groups longer. Maturing males help their fathers defend their groups against takeover attempts. Collective defense is crucial to males' success because single males are unable to defend groups against incursions by rival males.

This situation leads to a kind of arms race, as migrating males also form coalitions and cooperate in efforts to evict residents. After they have established residence, males collectively defend the group against incursions by extragroup males. Cooperation among males is beneficial since it helps deter rivals. But it also involves clear fitness costs because, as behavioral and genetic data have demonstrated, only one male succeeds in siring offspring within the group. Not surprisingly, kinship influences the duration and stability of male coalitions. Coalitions that are made up of related males last nearly four times as long as coalitions composed of unrelated males. Coalitions composed of related males are also less likely to experience rank reversals. In this case, the costs of cooperation may be balanced by gains in inclusive fitness.

Reciprocal Altruism

Altruism can also evolve if altruistic acts are reciprocated.

The theory of **reciprocal altruism** relies on the basic idea that altruism among individuals can evolve if altruistic behavior is balanced between partners (pairs of interacting individuals) over time. In reciprocal relationships, individuals take turns being actor and recipient—giving and receiving the benefits of altruism (Figure 8.16). Reciprocal altruism is favored because over time the participants in reciprocal acts obtain benefits that outweigh the costs of their actions. This theory was first formulated by Robert Trivers of Rutgers University, and later amplified and formalized by others.

Three conditions occurring together favor the development of reciprocal altruism: individuals 1) must have an opportunity to interact often, 2) must have the ability to keep track of support given and received, and 3) must provide support

FIGURE 8.16

Two old male chimpanzees groom each other. Reciprocity can involve taking turns or interacting simultaneously. Male chimpanzees remain in their natal communities throughout their lives and develop close bonds with one another.

How Do We Study Primates?

Habituation and Recognition

At the beginning of any study the field primatologist faces a number of problems. He or she must locate subjects, habituate them to the presence of humans without disrupting their natural behavior, distinguish one group from another, and, if possible, learn to recognize animals individually. These early stages of fieldwork often represent a low point in morale and productivity. To begin with, habituating monkeys to close-range observation is extremely frustrating and dull. Little can be accomplished if your subjects vanish whenever you approach, although good work on vegetation, fecal samples, and warning cries has been conducted during this period. When the observer is finally lucky enough to get close to the subjects, attempts to recognize animals individually can create further frustration and anguish. In some studies, of course, the difficulties of individual recognition have been overcome by capturing and marking animals, while observers of species such as chimpanzees and gorillas report that their animals can be recognized individually as easily as humans, even after only one or two days' observation. For the most part, however, primatologists give thanks for cuts on the ear, misshapen noses, wounds that leave scars, and other abnormalities that facilitate recognition until, after months of frustration, individual monkeys are distinguished almost as easily as individual humans.

Observational Sampling

Once a group of primates has been habituated and the animals can be recognized individually, the observer is ready to begin sampling their behavior. . . . Underlying such observations is a fundamental principle: one cannot record everything. Even in small groups, social behavior or feeding can occur in rapid sequences, and the observer will often want to gather carefully limited data on the durations of events, intervals between events, or sequences of events. From a large population of many behavioral acts, the observer must therefore sample a small proportion and design rules for sampling to yield data that comprise a representative, unbiased subset of all the events that have occurred.

. . . We briefly describe here perhaps the three most common methods of observational data collection.

Instantaneous Sampling. This method, also called point sampling, is used to record one or more classes of behavior at a predetermined instant in time. For example, suppose an observer wanted to determine whether individuals in a group of black-and-white colobus monkeys showed a particular pattern of feeding and resting throughout the day, perhaps feeding early in the morning, late in the afternoon, and resting at noontime. A test of this hypothesis using instantaneous sampling would proceed as follows. First, decide the time interval at which monkeys will be sampled, for example, one individual every minute. Second, produce a randomized list of the individual monkeys to be sampled. Third, beginning at the earliest light of dawn, locate the first monkey listed and record its activity at the instant that the minute changes or over a predetermined, brief interval (e.g., 5 sec) beginning on the minute. Then, repeating this procedure for the next monkey on the list, continue until all subjects have been sampled equally often for a specified length of time.

If this protocol is repeated for a number of days, and if sampling sessions are varied so that they include an equal number of sessions from dawn to dusk, the observer will find it fairly easy to compare, for each individual, the proportion of time spent in different activities at different times of day. Instantaneous sampling is often useful because it allows a large number of individuals to be sampled daily, and it minimizes the length of time between successive samples on the same individual. . . . The value of instantaneous samples is limited, however, because it provides no information on sequences of interactions or their duration.

Focal Animal Sampling. Suppose one wanted to test the hypothesis that genetically related individuals were more likely than others to groom within 1 minute after an aggressive interaction. Instantaneous sampling would clearly be inadequate to address this question, but an appropriate method could be devised if the observer simply increased the duration of time during which data were recorded. For instance, an observer might locate the first animal on the list and then follow that individual, recording all its activities, not for an instant but for some predetermined time, say 5 minutes, 10 minutes, 1 hour, or even a day. Such long-term follows of a preselected individual are an example of focal animal sampling, perhaps the most widely used method in the study of primate behavioral ecology. Focal animal sampling is an extremely versatile technique because the data it produces can be used to answer a variety of questions about frequencies (how often a behavior occurs), rates (how often a behavior occurs per unit time), sequences (how often behavior A follows behavior B), and durations (how long a behavior typically lasts). . . . One weakness of focal animal sampling, however, results from the time devoted to each sample. It is rarely possible to sample many animals on a given day, and considerable time may elapse between successive samples on the same individual. . . .

Many primate field projects benefit from the assistance of skilled observers like Mr. Mokupi Mokupi, who studies baboons in the Okavango Delta of Botswana.

Ad libitum Sampling. A variety of conditions may conspire to make regular, ordered sampling of behavior by either instantaneous or focal animal sampling impossible. In dense vegetation, for example, it may prove extremely difficult to find "the next subject" on an observer's list, and all primatologists are familiar with the mysterious process that causes subjects to disappear from sight shortly after a focal sample on them has begun. More important, some extremely significant behaviors such as copulation may be so rare that most instances would go unrecorded even by regular observation. To overcome these limitations, a variety of ad libitum sampling techniques, each designed to supplement data gathered by more systematic methods, have been developed by different observers. Each method of ad-lib sampling has its own assets and limitations, and each reflects the different problems posed by work on different species. . . .

Field Experiments

Field experiments have always played a major role in research on birds and nonprimate mammals, but until recently such experiments were rarely used by primatologists, for a number of different reasons. Most important, the same logistical factors that prevent observation at close range can discourage any thought of experimentation in the field. Accounts of the abnormal behavior that results when primates interact with humans in zoos or wild animal parks have encouraged the belief that behavior will quickly be distorted if the animals interact with their observers in any way. These considerations are certainly understandable and important, but it has also become clear during the past ten years that, with suitable precautions, well-controlled field experiments can be conducted on primates without distorting their behavior and that, as with other species, such experiments can provide new insights into behavior that cannot be obtained through observation alone.

Among primatologists, Hans Kummer and his colleagues were pioneers in the integration of observational and experimental field techniques. . . . Beginning with detailed observational data on hamadryas baboons, these investigators formulated hypotheses about the mechanisms underlying food gathering, partner preferences, and social structure. They then tested such hypotheses by experimentally manipulating food availability or social competition, either among their free-ranging subjects or in large enclosures. . . .

A second group of primatologists, following a long tradition of field experiments on bird vocalizations, has used portable tape recorders and loudspeakers to conduct playback experiments on free-ranging primates.... Such work was originally designed to provide new information on the use of vocalizations, and in this area it has made important contributions.... Perhaps more important, however, vocal playbacks offer an opportunity to test hypotheses about an animal's social or ecological knowledge because they allow an observer to mimic the presence of certain individuals or the occurrence of certain events under specified conditions.

SOURCE (excluding photograph): From pp. 5–8 in B. B. Smuts, D. L. Cheney, R. M. Seyfarth, R.W. Wrangham, T. T.Struhsaker, 1987, The study of primate societies, pp. 1–8 in Primate Societies, ed. by B. B. Smuts et al., University of Chicago Press, Chicago. Copyright (1987) by The University of Chicago.

only to those who help them. The first condition is necessary so that individuals will have the opportunity for their own altruism to be reciprocated. The second condition is important so that individuals can monitor and balance altruism given and received from particular partners. The third condition produces the nonrandom interaction necessary for the evolution of altruism. If individuals are unrelated, initial interactions will be randomly distributed to altruists and nonaltruists. However, reciprocators will quickly stop helping those who do not help in return, while they will continue to help those who do. Thus, as in the case of kin selection, reciprocal altruism can be favored by natural selection because altruists receive a disproportionate share of the benefits of altruistic acts. Note that altruistic acts need not be exchanged in kind; it is possible for one form of altruism (such as grooming) to be exchanged for another form of altruism (such as coalitionary support).

In primates, the conditions for the evolution of reciprocal altruism probably are satisfied often, and there is some evidence that it occurs.

Most primates live in social groups that are fairly stable, and they can recognize all of the members of their groups. We do not know whether primates have the cognitive capacity to keep track of support given and received from various partners, but we know they are very intelligent and can solve complex social problems. Thus, it seems likely that primates have both the opportunity and the capacity to practice reciprocal altruism.

Reciprocal grooming and coalitionary support have been observed in several species of macaques, baboons, vervet monkeys, and chimpanzees. In some cases, grooming and support are reciprocated in kind, while in other cases monkeys seem to exchange grooming for support. Among male chimpanzees, social bonds seem to be based on reciprocal exchanges in many different currencies. Thus, John Mitani and David Watts have found that male chimpanzees at Ngogo, a site in the Kibale Forest of Uganda, share meat selectively with males who share meat with them and with males who regularly support them in agonistic interactions. Moreover, males who hunt together also tend to groom one another selectively, support one another, and participate in border patrols together. Interestingly, close associates are not maternal kin, suggesting that males' relationships are based on reciprocity, not kinship.

These correlational findings are consistent with predictions derived from the theory of reciprocal altruism, but they do not demonstrate that altruism is contingent

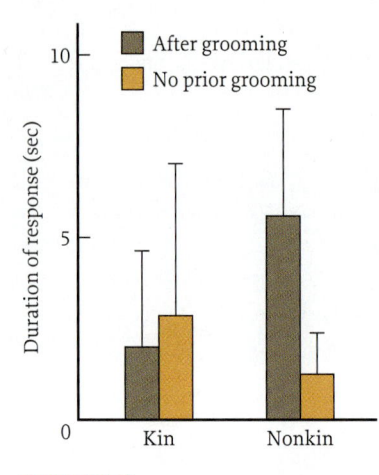

FIGURE 8.17

Vervet monkeys responded more strongly to recruitment calls played from a hidden speaker if the caller had previously groomed them than if the caller had not.

on reciprocation. Data from a number of recent studies suggest that primates do keep track of these contingencies, at least over short periods.

Robert Seyfarth and Dorothy Cheney conducted the first study to examine the contingent nature of altruistic exchanges. Like most monkeys, vervets spend much of their free time grooming. Vervets also form coalitions and use specific vocalizations to recruit support. In this experiment, Seyfarth and Cheney played tape-recorded recruitment calls to individuals in two different situations. Vervet A's recruitment call was played to vervet B from a hidden speaker 1) after A had groomed B and 2) after a fixed period of time in which A and B had not groomed. It was hypothesized that if grooming were associated with support in the future, then B should respond most strongly to A's recruitment call after being groomed. The vervets did just that (Figure 8.17).

Frans de Waal of Emory University designed another study, which focused on the relationship between grooming and food sharing in a well-established group of captive chimpanzees. At various times over a three-year period, a group of chimpanzees was provided with compact bundles of leaves. Although these bundles could be monopolized by a single individual, possessors of these bundles often allowed other individuals to share some of their booty. Before each provisioning session, de Waal monitored grooming interactions among group members. He found that the possessors were more generous to individuals who had recently groomed them than they were toward other group members (Figure 8.18). Moreover, possessors were less likely to actively resist efforts to obtain food if their solicitors had recently groomed them than if they had not groomed them.

The number of well-documented examples of reciprocal altruism in nonhuman primates is still small. It is possible that reciprocity is actually uncommon in nature. However, reciprocal altruism may occur more often than it is detected by observers. Since altruism is potentially reciprocated in different currencies (grooming for support, predator defense for food, and so on), and since the actual costs and benefits associated with specific behaviors are almost impossible to quantify, it is very difficult to establish whether altruism is actually reciprocated.

FIGURE 8.18

Chimpanzees were more successful in obtaining food from animals whom they had previously groomed (left bar) than from animals who had been groomed by others, or not groomed at all.

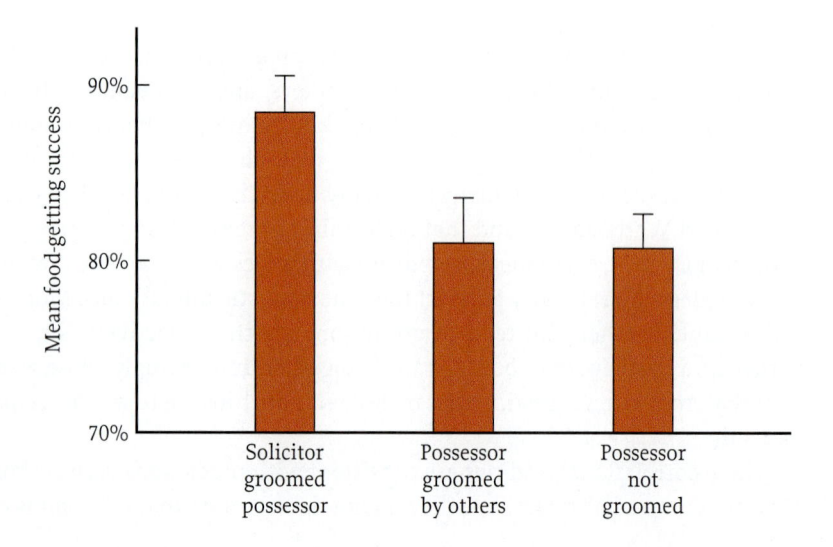

Further Reading

Chapais, B. 1995. Alliances as a means of competition in primates: evolutionary, developmental, and cognitive aspects. *Yearbook of Physical Anthropology* 38:115–136.

Dugatkin, L. A. 1997. *Cooperation among Animals.* Oxford University Press, Oxford.

Harcourt, A. H., and F. B. M. de Waal, eds. 1992. *Coalitions and Alliances in Humans and Other Animals.* Oxford University Press, Oxford.

Kapsalis, E., and C. M. Berman. 1996. Models of affiliative relationships among free-ranging rhesus monkeys *(Macaca mulatta).* I. Criteria for kinship. *Behaviour* 133:1209–1234.

Krebs, J. R., and N. B. Davies. 1993. *An Introduction to Behavioural Ecology.* Sinauer Associates, Sunderland, Mass., chap. 11.

Seyfarth, R. M., and D. L. Cheney. 1984. Grooming, alliances, and reciprocal altruism in vervet monkeys. *Nature* 308:541–543.

Seyfarth, R. M., and D. L. Cheney. 1988. Empirical tests of reciprocity theory: problems in assessment. *Ethology and Sociobiology* 9:181–188.

Waal, F. B. M. de. 1997. The chimpanzees's service economy: food for grooming. *Evolution and Human Behavior* 18:375–386.

Study Questions

1. Consider the kinship diagram shown here. What is the kinship relationship (for example, mother, aunt, or cousin) and degree of relatedness (such as 0.5 or 0.25) for each pair of individuals?

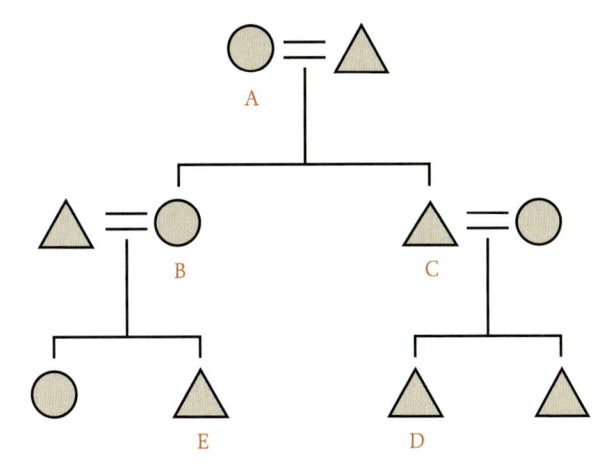

2. In biological terms, what is the difference between the following situations: (a) A male monkey sitting high in a tree gives alarm calls when he sees a lion at a distance. (b) A female monkey abandons a desirable food patch when she is approached by another female.

3. In documentaries about animal behavior, animals are often said to do things "for the good of the species." For example, when low-ranking animals do not reproduce, they are said to be forgoing reproduction in order to prevent the population from becoming too numerous and exhausting their resource base. What is wrong with this line of reasoning?

4. Why is some sort of nonrandom interaction among altruists necessary for altruism to be maintained?
5. Suppose that primates are not able to recognize paternal kin, as many primatologists have assumed. What would that tell you about how natural selection produces adaptations?
6. Data from a number of studies indicate that primates are more likely to behave altruistically toward kin than nonkin. However, many of the same studies show that rates of aggression toward kin and nonkin are basically the same. How does this fit with what you have learned about kin selection? Why are monkeys as likely to fight with kin as nonkin?
7. There are relatively few good examples of reciprocal altruism in nature. Why is reciprocal altruism uncommon? Why might we expect reciprocal altruism to be more common among primates than among other kinds of animals?
8. Some of the evidence for reciprocal altruism comes from correlational studies which show that primates are most likely to groom, support, or share food with the animals that are most likely to groom, support, or share food with them. What do correlational studies like these tell you about the causal forced involved in the distribution of these behaviors?
9. In Seyfarth and Cheney's experiment, they found that vervet monkeys tended to respond more strongly to the calls of the animals that had groomed them earlier in the day than to the calls of animals that had not groomed them. However, this effect held only for unrelated animals, not for kin. The vervets responded as strongly to the calls of grooming relatives as to those of nongrooming relatives. How might we explain kinship's influence on these results?
10. In addition to kin selection and reciprocal altruism, a third mechanism leading to nonrandom interaction of altruists has been suggested. Suppose altruists have an easily detected phenotypic trait, perhaps a green beard. Then altruists could use the following rule: "Do altruistic acts only for individuals who have green beards." Once the allele became common, most individuals carrying green beards would not be related to one another, so this would not be a form of kin selection. However, there is a subtle flaw in this reasoning. Assuming that the genes controlling beard color are at different genetic loci than the genes controlling altruistic behavior, explain why green beards would not evolve.

CHAPTER 9

Primate Intelligence

WHAT IS INTELLIGENCE?
WHY ARE PRIMATES SO SMART?
  HYPOTHESES EXPLAINING PRIMATE INTELLIGENCE
  TESTING MODELS OF THE EVOLUTION OF INTELLIGENCE
    KNOWLEDGE ABOUT THE ECOLOGICAL AND SOCIAL DOMAINS
    THE ECOLOGICAL DOMAIN
    SOCIAL KNOWLEDGE
    THEORY OF MIND
    THE GREAT APE PROBLEM
  THE VALUE OF STUDYING PRIMATE BEHAVIOR

What Is Intelligence?

Compendiums of the names for multitudes of animals invariably include some terms that are familiar (such as "flock of sheep" and "herd of buffalo"), some that are quaint (for example, "sounder of wild hogs"), and some that are particularly apt ("gaggle of geese"). A lesser-known entry in the last category is a "shrewdness of apes." Although there may have been little empirical evidence to justify this label when it was coined, we now know that it is an accurate assessment. Primates are unusual, if not unique, in the relatively large size of their brains and the complexity of their behavior. Monkeys and apes have larger brains in relation to their body size than do members of any other taxonomic group (Figure 9.1). Humans, of course, carry this evolutionary trend to an even greater extreme.

Large brains come at a substantial cost because brain tissue is extremely expensive to maintain. Our brains account for approximately 2% of our total body weight, but they consume about 20% of our metabolic energy. Natural selection does not maintain costly features unless they confer important adaptive advantages on organisms. Thus, one of the central questions of primate evolution is: why has evolution made

FIGURE 9.1

This graph plots the average value for brain and body weights for marsupials and for 10 orders of placental mammals. The red line represents the line that best fits the data for placental mammals. There is generally a uniform relationship between brain weight and body weight, with brain weight (in milligrams) being approximately equal to body weight (in grams) raised to the 0.75 power. Primates deviate from this best-fit line, which means that they have relatively large brains for their bodies. B = bats, C = carnivores, D = whales and dolphins, E = edentates, I = insectivores and tree shrews, L = rabbits and hares, M = marsupials, P = primates, R = rodents, S = seals and sea lions, U = hoofed mammals.

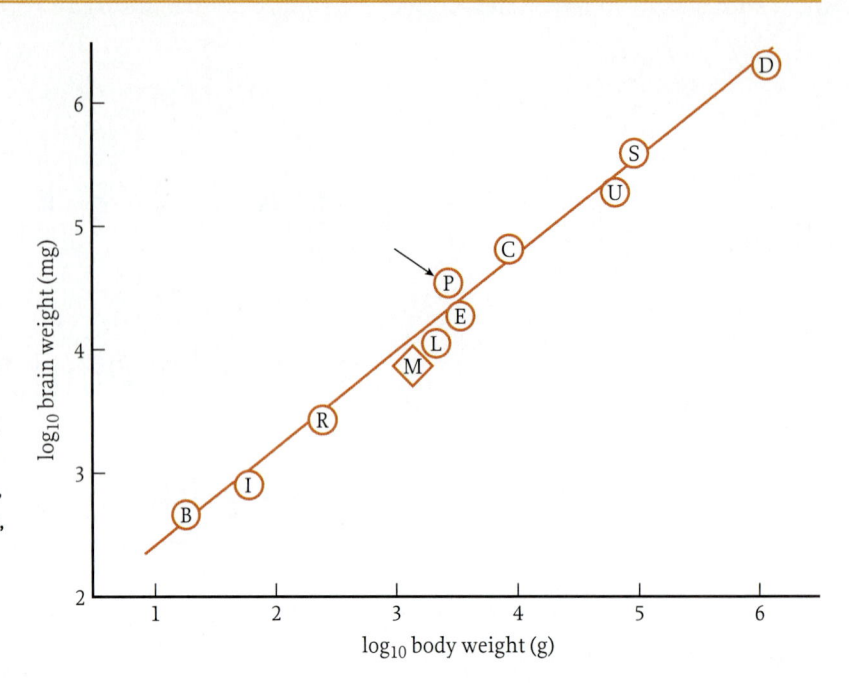

primates so smart? Understanding the nature and causes of the cognitive abilities of nonhuman primates will help us to answer this question. We begin this chapter by considering the selective factors favoring intelligence and behavioral complexity within the order Primates, and then examine evidence regarding the extent of cognitive complexity in nonhuman primates.

Flexible problem-solving ability is a primary component of intelligence.

As you know, intelligence is notoriously difficult to define. Many researchers avoid the term entirely, referring instead to specific cognitive capacities or adaptations that humans possess. It is even more difficult to measure intelligence or evaluate the nature of cognitive adaptations in members of other species, partly because we know so little about their mental processes. Considerable ink has been spilled over the question of whether other animals have mental representations, thoughts, or consciousness.

Although we might not be able to agree on a precise definition of intelligence, most of us would probably agree that one fundamental component of intelligence is the ability to solve problems in complex situations by flexible means. It is important to emphasize that intelligence and evolutionary success are not necessarily synonymous. If they were, then we would have no basis for distinguishing between a bird that recalls thousands of locations where it has stored seeds for the winter, a whale that finds its way from a beach in southern Argentina to summer fishing grounds in the Arctic Circle, a wildebeest female that manages to protect her calf from a hungry pack of hyenas, and a chimpanzee that leads his companions away from a hidden cache of prized food and returns to eat it in privacy. While the first three traits—a capacious memory, impressive navigational abilities, and effective parenting strategies—clearly contribute to the fitness of these organisms, they may be based on fixed stimuli or invariant rules that regulate behavior. Intelligence implies something more about the flexibility of the means used to solve problems. Therefore, our definition emphasizes the ability to cope with complexity and to incorporate novel solutions into existing behavioral repertoires (Figure 9.2).

FIGURE 9.2

Behaviors that increase an animal's fitness are not necessarily evidence of intelligence. For example, (a) whales—like many other animals—migrate long distances to find food, and (b) a female wildebeest will defend her calf against attack. Yet neither navigational abilities nor parenting behaviors require the kind of flexibility necessary to solve novel problems—a key component of intelligence.

Why Are Primates So Smart?

There is now considerable debate about the primary factors that favored the evolution of relatively large brains and enhanced cognitive capacities among nonhuman primates. Some researchers argue that ecological factors associated with locating and processing inaccessible food items are principally responsible for the evolution of intelligence among primates. Others have suggested that social demands associated with life in large, stable groups provided the primary selective force favoring cognitive complexity and intelligence among primates. We will review the rationales for each of these positions and then consider the empirical data that bear on each.

HYPOTHESES EXPLAINING PRIMATE INTELLIGENCE

Some primatologists believe that primate intelligence evolved to solve ecological problems.

University of California, Berkeley, anthropologist Katherine Milton believes that the challenges associated with exploiting short-lived and patchily distributed foods may have favored greater cognitive abilities among primates. As we discussed in Chapter 6, fruit trees in tropical forests are patchily distributed and bear ripe fruit for relatively short periods of time. Once an animal has located a food source, it will become a dependable, but seasonal, resource over the lifetime of the individual (Figure 9.3). It would be advantageous for primates to form a detailed mental map of the sites at which fruit may be found and to be able to find their way from one food source directly to another. Milton argues that individuals who can plan routes efficiently, anticipate the availability and quality of various food resources, and keep track of changing conditions of food sources may be at a distinct advantage.

Katherine Gibson of the University of Texas and Sue Parker of the State University of California at Sonoma offer a second ecological hypothesis. They believe that natural selection has favored enhanced cognitive capacities among primates because many foods that primates consume re-

FIGURE 9.3

A muriqui feeds on fruit. A reliance on patchily distributed foods like fruit may place a premium on cognitive skills. (Photograph courtesy of Karen Strier.)

FIGURE 9.4

Primates sometimes exploit foods that are difficult to extract. Here, (a) a male chimpanzee pokes a long twig into a hole in a termite mound and extracts termites, and (b) a capuchin monkey punctures an eggshell and extracts the contents. (b, Photograph courtesy of Susan Perry.)

(a)   (b)

quire considerable skill to process. For example, some primates eat hard-shelled nuts that must be cracked open with stones or smashed against a tree trunk. Others dig up roots and tubers; extract insect larvae from bark, wood pith, or dung; and crack open pods to obtain seeds (Figure 9.4). These kinds of foods, which Gibson and Parker label **extractive foods,** are valuable because they tend to be rich sources of protein and energy. However, extractive foods require complicated, carefully coordinated techniques to process. For example, to feed on the pith of wild celery plants, mountain gorillas must first break the stem into manageable pieces, then peel the outer layers of the stalk away with their teeth, and finally pick out edible bits of the pith with their fingers. To master these skills primates must be more intelligent than other animals.

Other scientists believe that primate intelligence evolved in order to solve social challenges.

In contrast to these ecological models, a number of primatologists believe that social challenges have provided the most important selective factor favoring the evolution of intelligence in primates (Figure 9.5). As we have seen, life in social groups means that animals face competition and experience conflict. At the same time, social life also provides opportunities for affiliation, cooperation, nepotism, and reciprocity. This view has come to be known as the **social intelligence hypothesis.**

FIGURE 9.5

Living in complex social groups may have been the selective factor favoring large brains and intelligence among primates.

TESTING MODELS OF THE EVOLUTION OF INTELLIGENCE

Comparative tests of the social and ecological hypotheses for the evolution of intelligence can be made by examining the relationship between the relative size of the brain's neocortex and the extent of social and ecological complexity across primate species.

Robin Dunbar, a primatologist at the University of Liverpool, pointed out that these models of the evolution of intelligence in primates generate specific predictions about the patterning of variation in intelligence or cognitive abilities among living primates. For example, Parker and Gibson's idea can be tested in theory by determining whether species that process complex, packaged foods are more intelligent than species that feed on simpler foods. However, a major obstacle to making such tests is the lack of reliable operational criteria to assess and compare cognitive abilities, ecological features, and social complexity among primate species.

The assessment of cognitive abilities is particularly problematic since we do not know exactly how the brain processes information or solves problems. Dunbar believes that the volume of the brain's neocortex provides a useful anatomical measure of cognitive capacity in primates for two reasons. First, the evolutionary changes that have occurred in the primate brain have mainly involved changes in the size and structure of the forebrain, particularly the **neocortex** (Figure 9.6). Second, the

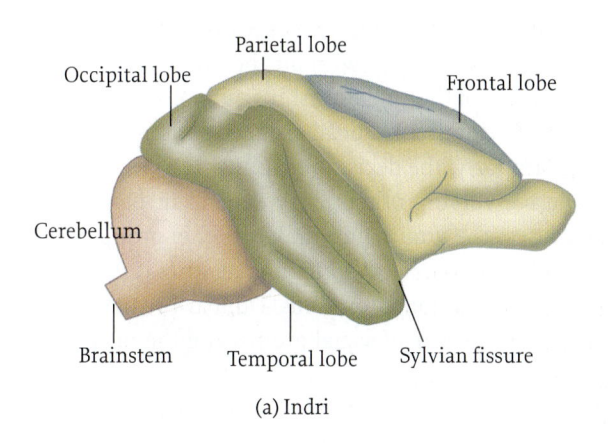

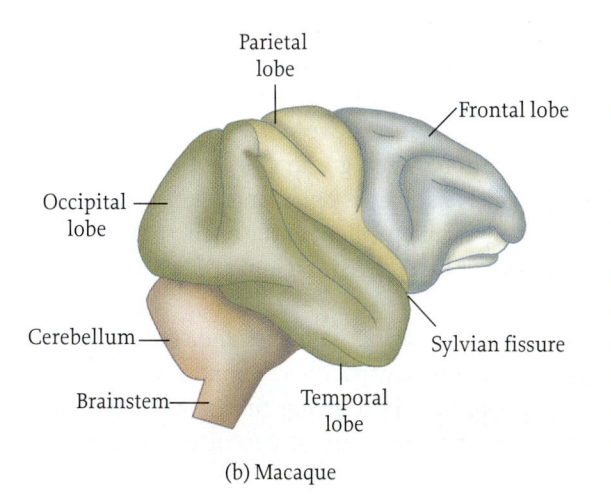

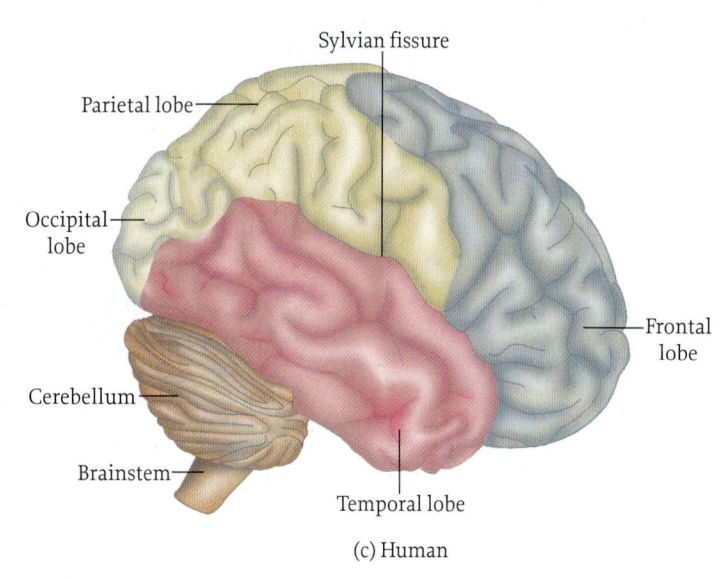

FIGURE 9.6

These figures depict the brains of (a) a large prosimian, the indri; (b) a macaque; and (c) a human. There are three main components of the brain: the hindbrain, which contains the cerebellum and medulla (part of the brainstem); the midbrain, which contains the optic lobe (not visible here); and the forebrain, which contains the cerebrum. The cerebrum is divided into four main lobes: occipital, parietal, frontal, and temporal. The forebrain is greatly expanded in primates and other mammals, and much of the gray matter (which is made up of cell bodies and synapses) is located on the outside of the cerebrum in a layer called the cerebral cortex. The neocortex is a component of the cerebral cortex, and in mammals the neocortex covers the surface of virtually the entire forebrain.

neocortex seems to be the thinking part of the brain, the part of the brain most closely associated with problem solving and behavioral flexibility. Therefore, Dunbar contends that the size of the neocortex and the **neocortex ratio,** the ratio of the size of the neocortex to the rest of the brain, can serve as a proxy for cognitive ability.

Both social and ecological hypotheses receive some support from comparative studies of the neocortex.

The ecological hypotheses for the evolution of intelligence predict that specific characteristics of the diet or the environment of particular primate species will be correlated with their cognitive abilities. For Milton's model, this means that smarter species should use more patchy food resources. Frugivores, which utilize a dispersed, patchy, and ephemeral (short-lived) food supply, would need greater cognitive skills than folivores, whose foods are more uniformly distributed. If the hypothesis is correct, we should find that species with a high proportion of fruit in the diet will have a higher neocortex ratio. It is also possible that the size of the home-range area represented by the mental map places the greatest demands on the brain. Frugivores typically have larger home ranges and longer day journeys than folivores. Thus, Milton's hypothesis predicts that neocortex ratio and home-range size or day-journey length will be positively correlated. Gibson's model predicts extractive foragers will have relatively larger neocortex ratios than will those who feed on more readily accessible foods.

The social intelligence hypothesis predicts that there will be a positive correlation between the complexity of social life and the neocortex ratio. Dunbar suggests that group size may be taken as a rough index of social complexity because primates in social groups need to recognize, associate, and interact with all the other members of their groups. Animals somehow keep track of their relationships with other members of their groups, particularly when they participate in social interactions that are regulated by nepotism or by reciprocity. Similarly, the decision to act aggressively or submissively, or to intervene in conflicts involving others, may depend in part on an individual's ability to remember or assess the dominance ranks of other group members. As groups grow larger, the number of pairs grows rapidly, making it considerably more difficult to keep track of social relationships. Thus, Dunbar predicts there should be a positive correlation between the size of social groups and the neocortex ratio.

Dunbar compiled data to test these predictions, gathering together data on group size, ecological parameters, and brain size from a variety of sources. He found neocortex ratio and neocortex size to be positively related to group size (Figure 9.7). That is, primates who live in large social groups tend to have larger neocortexes and greater neocortex ratios than those who live in smaller social groups. More recent work by Rob Barton of the University of Durham has extended Dunbar's result using statistical methods that control for phylogenetic affinities between species (as described in Box 4.1). Barton confirmed the relationship between group size and neocortex size. He also discovered that frugivorous monkeys and apes have larger neocortexes than other species do, and that this relationship is independent of the effects of group size. There is no evidence that home-range size or extractive foraging is directly associated with neocortex enlargement.

These data provide support for the idea that social challenges and ecological parameters played a role in the evolution of cognitive capacities in nonhuman primates. However, there are two reasons to be cautious in drawing firm conclusions from these data. First, we cannot be sure that neocortex ratio or neocortex size are valid

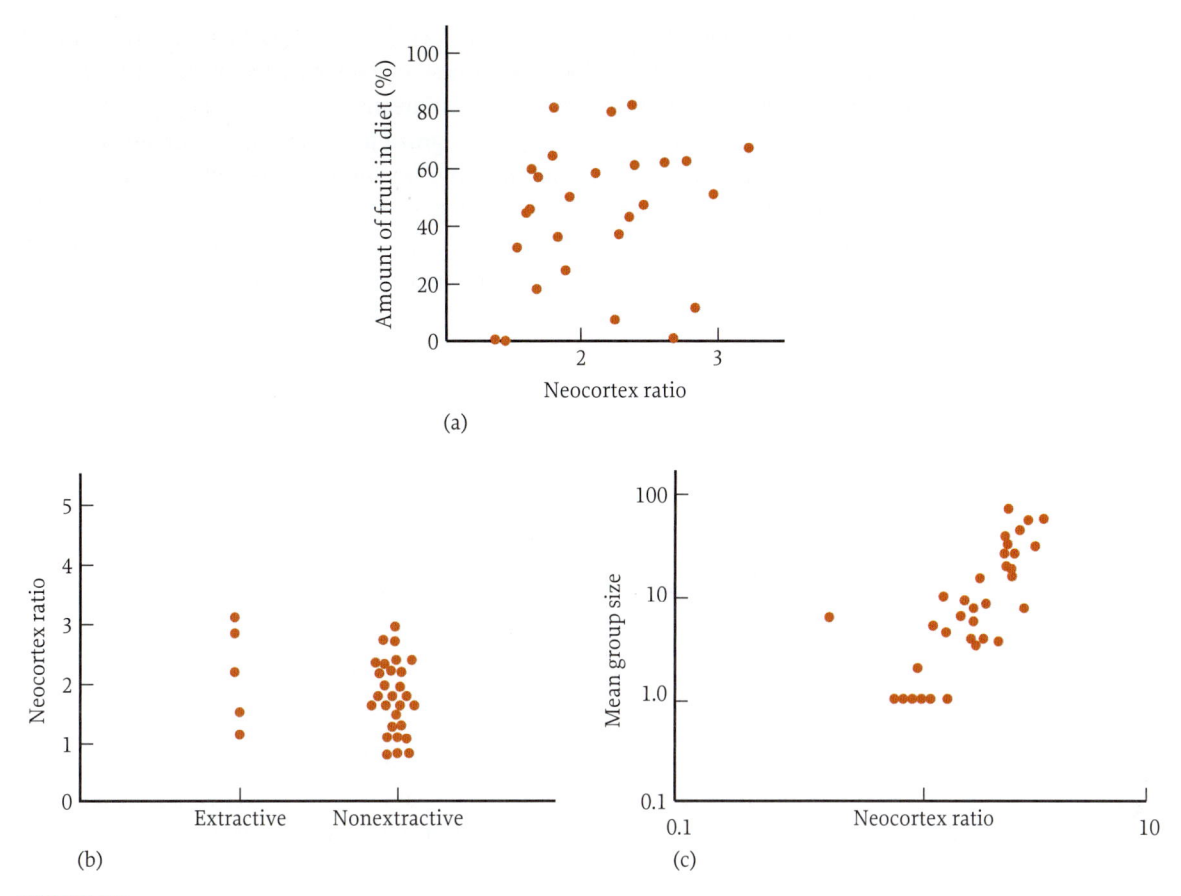

FIGURE 9.7

The neocortex ratio (the ratio of the volume of the neocortex to the volume of the rest of the brain) is plotted against (a) the amount of fruit in the diet, (b) extractive versus nonextractive foraging techniques, and (c) group size. Primates who live in large groups have larger neocortex ratios than those who live in smaller groups.

measures of cognitive ability because we know very little about how the brain processes information or shapes behavior. Second, it is always problematic to infer causal relationships from correlational data. The association between group size and neocortex ratio may be a spurious result that arises because both group size and neocortex ratio are causally related to a third variable that has not yet been identified. Thus, we now turn to observational and experimental evidence about what monkeys and apes know about their world.

Knowledge about the Ecological and Social Domains

If ecological pressures have favored the evolution of intelligence, then primates should be adept at solving ecological problems. Similarly, if social pressures have favored the evolution of intelligence, then primates should be adept at solving social challenges.

Certain problems arise when we attempt to measure what monkeys actually know about their world. Unlike psychologists working with human subjects, we cannot ask monkeys what they are thinking. At the same time, we cannot simply infer what they know or what they understand from what they do.

To see why this might be problematic, consider the following example. An observer watching a female macaque notes that she does not respond when her infant screams. How can the observer determine whether this means that the mother is 1) unaware that her infant is in distress, 2) unable to recognize her own infant's scream, or 3) reluctant to intervene against a higher-ranking female than herself? To distinguish among these three possibilities, the observer must be able to demonstrate that mothers can distinguish between calls given in different contexts, that mothers can recognize their own infants' calls and differentiate them from the calls of other infants, and that mothers can assess the nature of the risks their infants face. As daunting as these tasks seem, researchers working in field and laboratory situations have developed methods for answering these kinds of questions. As a result of this innovative experimental work, we are beginning to get some idea of what monkeys know about their world.

THE ECOLOGICAL DOMAIN

Monkeys and apes know a lot about their environments.

Monkeys are excellent naturalists. They select appropriate food items, avoid eating foods laced with toxins, know the location of food resources within their home ranges, move efficiently from one feeding site to the next, and evade predators. Although these skills are clearly related to survival and reproductive success, there is a general assumption that these kinds of ecological tasks are less cognitively demanding than social tasks. But it is possible that environmental knowledge is more complicated than we think, perhaps because we are accustomed to buying our food in grocery stores and using maps and compasses to find our way in the wilderness. Consider, for example, the tasks primates must face living in the forests of East Kalimantan, where there are more than 700 species of fruiting plants. These animals must distinguish which fruits are edible and which are not, track the status of fruit crops as they ripen, monitor the availability of fruits at different sites, make decisions about when to remain in a food patch and when to switch to a new one, and so on.

Macaques know a lot about the characteristics of the foods they eat.

Charles Menzel of Georgia State University has designed a series of experiments to find out what macaques know about the foods they eat. For example, he has compared the responses of free-ranging Japanese macaques finding indigenous foods and provisioned foods in their home ranges. When they find akebi fruits on the ground when these fruits would not normally be present, the monkeys immediately look up into trees for additional fruits. When they find chocolate on the ground, the macaques search the ground for more chocolate in the immediate vicinity but don't look elsewhere. Apparently, these monkeys know that fruit grows on trees but candy doesn't. When monkeys are presented with two similar unfamiliar objects, a red rubber ball with a black band and a red plastic tomato, they focus their attention on the plastic tomato, apparently orienting to generic visual features that create resemblances to familiar foods. These simple experiments suggest that monkeys have considerable, subtle knowledge about the properties of familiar food items.

Monkeys and apes construct cognitive maps of their home ranges. These maps enable them to move efficiently from one food source to another.

Perhaps the most convincing evidence of the complexity of primates' ecological knowledge comes from studies of what monkeys know about the location of food and how they move efficiently from one feeding site to the next. The mental representation of the location, availability, and quality of things in the environment is called a **cognitive map.** Cognitive maps allow animals to work out efficient routes from one site to another, saving them both time and energy. To see why cognitive maps are so useful, consider the way you navigate your own environment. You begin at home, perhaps rushing to your first class of the day. Then you might head to the library to study, or to the cafeteria for lunch. Later, you might have another class at the other end of campus. At the end of the day, you might stop at the gym to work out. When you were new to campus, you probably took the same routes every day and were afraid to stray from familiar paths for fear of getting hopelessly lost in the crowded maze of buildings. However, as time passed, you discovered new routes that shortened your walk from one place to another, and you felt more confident in your ability to get around campus. You could do this because of an ability to visualize where things were and an ability to plan efficient routes. You developed a cognitive map of your home range. Many other primates apparently have the same kinds of abilities.

Paul Garber of the University of Illinois at Urbana-Champaign has studied the foraging behavior of tamarins in the forests of Peru. Some of the plant species that tamarins feed upon are highly synchronous. This means that if one individual of a particular species is fruiting or producing gum, others of the same species are doing the same thing. Thus, when tamarins have exhausted the resources of one feeding tree, they can expect to find more food at other trees of the same species. Garber found that 70% of the time, tamarin groups moved directly from one feeding tree to the nearest tree of the same species that they had not yet depleted. These tiny monkeys apparently know the location of hundreds of food trees and can remember for days or weeks when they last fed at particular sites.

Charles Janson of the State University of New York at Stony Brook has conducted a series of field experiments to examine capuchin monkeys' knowledge of the location and quality of their food resources. Janson constructed special feeding platforms in the forest and baited the platforms with tangerines. He conducted the experiments during the winter months when virtually no other fruit was available to the monkeys. Janson varied the amount of food available on each platform, so some sites always contained more fruit than others, and no site was baited more than once per day. By following the monkeys' movements through the forest, he was able to examine their knowledge of the locations and quality of these artificial food sites. Janson found that monkeys moved to closer platforms more often than they moved to more distant ones. The capuchins preferred sites where they might expect to find more fruit over sites where they might expect to find less fruit, and they avoided visiting sites that had been depleted within the last 24 hours. These data suggest that monkeys were able to remember the location of their fruit patches, to assess the amount of fruit that would be available at each site, and to evaluate the likelihood that the crop would be renewed since their last visit.

Taken together, these data suggest that all monkeys create good cognitive maps of their home ranges and have detailed knowledge of their foods' characteristics. Advocates of the social intelligence hypothesis downplay this evidence because these skills do not necessarily distinguish primates from other kinds of animals. They point out that rodents, birds, fish, and perhaps even insects have similar skills. However, those who emphasize the complexity of ecological knowledge also contend

FIGURE 9.8

In many primate species, all group members take an active interest in infants. Here a female baboon greets a newborn infant. Evidence from playback experiments, laboratory experiments, and naturalistic observations suggest that monkeys know something about the relationships among group members.

that we don't yet understand what primates know about their environment, making it premature to conclude that other animals know as much about the nonsocial world as primates do.

SOCIAL KNOWLEDGE   One of the most striking things about primates is the interest they take in one another. Newborns are greeted and inspected with interest (Figure 9.8). Adult females are sniffed and visually inspected regularly during their estrous cycles. When a fight breaks out, other members of the group watch attentively. As we have seen in previous chapters, monkeys know a considerable amount about their own relationships to other group members. A growing body of evidence suggests that monkeys also have some knowledge of the nature of relationships among other individuals, or **third-party relationships.** Monkeys' knowledge of social relationships may enable them to form effective coalitions, practice deception, and manipulate other group members for political purposes.

There is evidence that monkeys and apes know something about the nature of kinship relationships among other members of their groups.

One of the first indications that monkeys understand the nature of other individuals' kinship relationships came from a playback experiment on vervet monkeys conducted by Dorothy Cheney and Robert Seyfarth in Amboseli National Park. Several female vervets were played the tape-recorded scream of a juvenile vervet from a hidden speaker. When the call was played, the mother of the juvenile stared in the direction of the speaker longer than other females did. This suggests that mothers recognized the call of their own offspring. However, even before the mother reacted, other females in the vicinity looked directly at the juvenile's mother. This suggests that other females understood who the juvenile belonged to, and that they were aware that a special relationship existed between the mother and her offspring.

Additional evidence that monkeys may understand mother-offspring relationships comes from more controlled laboratory experiments conducted by Verena Dasser working with a captive group of long-tailed macaques in Zürich, Switzerland. In these experiments, a young female was the primary subject. In the first phase, the female was shown one slide of a particular mother-offspring pair and a second slide of two monkeys who were not related as mother and infant. All of the monkeys in the slides were members of the subject's social group and were well known to her. These trials were repeated until the female was able to select the mother-offspring pair consistently. In the second phase of the experiment, the female was again presented with two slides of group members, one depicting a mother-offspring pair and one depicting another pair of animals who were not mother and infant. However, in this phase of the experiment she was shown a different mother-offspring pair in each trial. If, *and only if,* she understood that there was a particular relationship between the mother and offspring that served as her model in training, she would select the new mother-offspring pair rather than the other pair of monkeys. The young female picked out the mother-offspring pair in nearly every trial. The monkey's performance in these experiments is particularly remarkable because the offspring in the mother-offspring pairs varied considerably in age, and in some cases the subjects had not seen the mother interacting with her child when it was an infant.

Monkeys may also have broader knowledge of kinship relationships. The evidence for this also comes from Cheney and Seyfarth's work on vervet monkeys. When monkeys are threatened or attacked, they often respond by threatening or attacking a lower-ranking individual that was not involved in the original incident, a phenomenon we call **redirected aggression.** Vervets selectively redirect aggression toward the maternal kin of the original aggressor. So, if female A threatens female B, then B threatens AA, a close relative of A. If monkeys were simply blowing off steam or venting their aggression, they would choose a target at random. Thus, the monkeys seem to know that certain individuals are somehow related.

Several lines of evidence suggest that monkeys understand the nature of rank relationships among other individuals.

Since kinship and dominance rank are major organizing principles in most primate groups, it makes sense to ask whether monkeys also understand third-party rank relationships. Although information about knowledge of rank relationships is limited, the evidence suggests that monkeys may understand the relative ranks of other group members.

The most direct evidence that monkeys understand third-party rank relationships comes from playback experiments conducted on baboons in the Okavango Delta of Botswana by Cheney, Seyfarth, and one of us (J. B. S.). In this group, dominance relationships were stable, and females never responded submissively toward lower-ranking females. In this experiment, females listened to a recording of a female's grunt followed by another female's submissive fear barks. Female baboons responded more strongly when they heard a higher-ranking female responding submissively to a lower-ranking female's grunt than when they heard a lower-ranking female responding submissively to a higher-ranking female's grunt. Thus, females were more attentive when they heard a sequence of calls that did not correspond to their knowledge of dominance rank relationships among other females. Control experiments excluded the possibility that females were reacting simply to the fact that they had not heard a particular sequence of calls before. The pattern of responses suggests that females knew the relative ranks of other females in their group and were particularly interested in the anomalous sequence of calls.

Participation in coalitions may draw on sophisticated cognitive abilities.

Even the simplest coalition is quite a complex interaction. When coalitions are formed, at least three individuals are involved, and several different kinds of interactions are going on simultaneously (Figure 9.9). Consider the case in which one monkey, the aggressor, attacks another monkey, the victim. The victim then solicits support from a third party, the ally, and the ally intervenes on behalf of the victim against the aggressor. The ally behaves altruistically toward the victim, giving support to the victim at some potential cost to itself. At the same time, however, the ally behaves aggressively toward the aggressor, imposing harm or energy costs on the aggressor. Thus, the ally simultaneously has a positive effect on the victim and a negative effect on the aggressor. Under these circumstances, decisions about whether or not to intervene in a particular dispute may be quite complicated. Consider a female who witnesses a dispute between two of her offspring. Should she intervene? If so,

FIGURE 9.9

Primates form coalitions that are more complicated than the coalitions of most other animals. (a) In a captive bonnet macaque group, members of opposing factions confront one another. (b) Two capuchins jointly threaten a third individual out of the picture. (b, Photograph courtesy of Susan Perry.)

(a) (b)

which of her offspring should she support? When a male bonnet macaque is solicited by a higher-ranking male against a male that frequently supports him, how should he respond? In each case, the ally must balance the benefits to the victim, the costs to the opponent, and the costs to itself (Figure 9.10).

Understanding of third-party relationships may be particularly useful in managing coalitions.

Given the great complexity of even the simplest coalitions, knowledge of third-party relationships may be extremely valuable because it enables individuals to predict how others will behave in certain situations. Thus, animals who understand the nature of third-party relationships may have a good idea about who will support them and who will intervene against them in confrontations with particular opponents, and they may also be able to tell which of their potential allies are likely to be most effective in coalitions against their opponents.

Primates seem to form more complex coalitions than other animals do.

Many animals form alliances in defense of their territories and their young. However, Andrew Harcourt at the University of California, Davis, has concluded that primates use alliances in different ways from other animals. Primates seem to assess differences in competitive ability among other members of their group and cultivate relationships with powerful individuals. For example, in many species, monkeys selectively groom higher-ranking individuals. In some of these species, grooming is exchanged for support in aggressive encounters, and individuals compete to groom high-ranking animals.

Primates may also attempt to manipulate the alliances among other group members. There are several accounts of chimpanzees attempting to prevent lower-ranking rivals from forming alliances with each other.

Deception provides further evidence of the sophistication of primate social knowledge.

FIGURE 9.10

Complex fitness calculations may be involved in decisions about whether to join a coalition. By helping the victim against the aggressor, the ally increases the fitness of the victim but decreases the fitness of the aggressor.

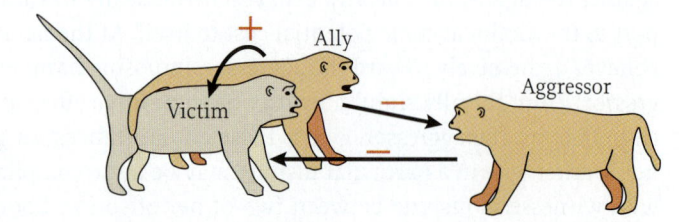

Box 9.1
Examples of Deception in Nonhuman Primates

The following examples of deception in nonhuman primates are adapted from A. Whiten and R. Byrne, 1988, The manipulation of attention in primate tactical deception, pp. 211–223 in *Machiavellian Intelligence*, ed. by R. Byrne and A. Whiten, Oxford University Press, Oxford.

Concealment Hans Kummer watched as an adult female hamadryas baboon spent 20 minutes inching toward a spot behind a rock. Once she reached this spot, she began to groom an adolescent male. This was an interaction not normally tolerated by the resident adult male.

Distraction Andrew Whiten and Richard Byrne observed a young male baboon attack a juvenile that screamed repeatedly. Screams are one means of recruiting support. As several adults came into view giving aggressive vocalizations, the attacker stood bipedally and stared intently into the distance as if a predator had been spotted. The adults paused and followed his gaze. Although no predator was detected, the aggressive interaction was terminated.

Creating a false image Frans de Waal, who conducted long-term observations of chimpanzees in the Arnhem Zoo (the Netherlands), described a male chimpanzee sitting with his back to a rival male. The male heard his rival give an aggressive vocalization and he grinned submissively. He used his fingers to push his retracted lips together over his teeth, altering his facial expression. He repeated this three times before the fear grin was eliminated. Then he turned to face his rival.

Manipulation of a conspecific using a social tool Robin Dunbar described one case in which an infant gelada baboon was unsuccessful in its efforts to nurse. The infant moved near the group's dominant male and then vocalized, hit the male's back, and pulled on the long mane of hair around his shoulders. The male ignored the infant, but the infant pulled on his hair again. The male then turned and struck at the infant. In the commotion, the mother looked up. When the infant approached her again, she allowed it to nurse.

Andrew Whiten and Richard Byrne of the University of St. Andrews in Scotland compiled and catalogued instances of deception in nonhuman primate species. They found evidence for a range of deceptive strategies employed by monkeys and apes (Box 9.1). Chimpanzees figure most prominently in Whiten and Byrne's compendium. Whiten and Byrne argue that these anecdotes are significant because they require considerable behavioral flexibility on the part of the actor. The acts are normal parts of the animal's behavioral repertoire, but the behaviors are used in unusual ways and contexts to achieve specific objectives that are beneficial to the actors. Thus, when a baboon suddenly looks intently toward the horizon in the midst of an aggressive contest, it uses a standard part of its behavioral repertoire. Such behavior generally means that a predator has been sighted, and is not a standard element in aggressive encounters. In this case, it has the effect of distracting the baboon's opponents and ending the conflict.

Deception among primates differs from the kinds of deception seen in other animals. Other animals feign injury to lure potential predators away from vulnerable young, they mimic the phenotypes of foul-tasting species, or they camouflage themselves to blend in with their surroundings (Figure 9.11), but there is not much flexibil-

FIGURE 9.11

Many animals, like this stick insect (a), camouflage themselves as a defense against predators (b). This is an effective form of deception, but it does not require sophisticated cognitive abilities.

ity in this deceptive behavior. Moreover, all of these forms of deception are principally directed at members of other species. Primates seem to be unusual in the flexible and tactical nature of their deception and in their capacity to deceive familiar members of their own species.

A number of researchers remain skeptical about the existence of deception among nonhuman primates and believe that each of the anecdotes catalogued by Whiten and Byrne has a simpler explanation. It is very difficult to prove that a given incident is the result of a conscious intention to deceive conspecifics. Behaviors like hiding from the dominant male while grooming a subordinate male, or distracting aggressors by feigning concern about a predator, may be random innovations that happen to work, not goal-directed strategies.

Because chimpanzees are able to manipulate other group members adeptly for strategic purposes, they may have mastered rudimentary political tactics.

In his influential book *Chimpanzee Politics*, Frans de Waal documents how male chimpanzees achieve and maintain high-ranking positions in a captive colony at the Arnhem Zoo in The Netherlands. He vividly describes power struggles among three rival males and their efforts to contain alliances that would threaten their own positions. The central players in this drama were Yeroen, Luit, and Nikkie. Yeroen and Luit were the two oldest males in the group, while Nikkie was considerably younger. For many years, Yeroen was the undisputed leader of the group, and his relationship with Luit was stable but never entirely relaxed. Then Luit abruptly began to challenge Yeroen's position, beginning a power struggle that continued for more than a year. When Luit's challenge began, Yeroen started to associate more often with the adult females in the group who invariably supported him in contests against Luit. Luit then began to harass the females when they associated with Yeroen, but he also groomed them assiduously at other times. Gradually the females began spending less and less time with Yeroen. During this period, Luit obtained support from the third male, Nikkie. Although Nikkie did not intervene directly against Yeroen on Luit's behalf, he did turn his attentions to those females who supported Yeroen. By harassing the females when they came to Yeroen's aid, he distracted them from supporting Yeroen. This in turn enabled Luit to intimidate Yeroen. After many months, Luit successfully established himself as the most dominant male and gained the support of most of the females who had formerly supported Yeroen. But further developments continued to alter the balance of power within the group when Yeroen and Nikkie formed a coalition that ultimately undermined Luit's position. Luit initially attempted to deter his

What Do Chimpanzees Think about Seeing?

[*Authors' note:* When we observe similar behavior in humans and other animals, we often assume that the behavior has a similar function and involves the same underlying processes. This assumption is often very useful in developing hypotheses about the origins of human behavior. However, there are some cases in which this assumption is probably incorrect. In this reading, careful laboratory studies reveal that there are fundamental differences in what chimpanzees and humans understand about seeing.]

During the first several years of our project, we devoted a considerable amount of time to trying to determine whether apes (like us) interpret the eyes as windows into the mind—and in particular, whether they have a concept of 'seeing'. Humans, of course, understand seeing as far more than just a geometric relation between eyes and objects and events in the world. We conceive of it as a subjective or psychological part of an experience that the other person is having: 'She *sees* me'. This level of understanding seeing may emerge quite early in human development—possibly by as young as two years of age.

Appreciating the idea that others 'see' is, in some sense, foundational to the entire question of theory of mind—at least with respect to our human understanding of the mind. After all, most of our social interactions begin with a determination of the attentional state of our communicative partners, and from that point forward we constantly monitor their attentional focus throughout the interaction. Nothing can disrupt a social interaction more quickly than realizing that someone is no longer looking at you. Furthermore, the appreciation that we *see* (and hence experience) each other is the glue that seems (to us, at least) to bind us to our communicative partners. True, we can arrive at this feeling of connection in other ways (for example, talking over the telephone does not make it possible to establish a sense of psychological connection to the other person). Nonetheless, the notion that the other person 'sees' is a basic, foundational assumption which, from our subjective point of view, seems to hold together most person-to-person interactions. . . .

Knowing That Others See You

Our first approach to asking our apes about 'seeing' had been to determine if they understood the psychological distinction between someone who could see them and someone who could not. We initially addressed this question by focusing on the natural begging gesture of chimpanzees. Chimpanzees use this gesture in several communicative contexts, including situations in which one ape is attempting to seek reassurance from another, or in cases where one ape is attempting to acquire food from another. Every day, our apes spontaneously use this gesture to request treats such as bananas, apples, sweet potatoes, onions, carrots, or even candy from their caretakers and trainers. For example, our chimpanzees frequently see us walk past their compound with some food, and immediately reach out, with their palms up (to 'request' the item, as it were) and then look into our eyes. Thus, this seemed like an ideal natural context in which to explore whether they appreciated that their gestures needed to be seen in order to be effective. . . .

We began by training the apes to enter the lab and gesture through a hole directly in front of a single, familiar experimenter who was either standing or sitting to their left or right. On every trial that they gestured through the hole directly in front of the experimenter, this person praised them and handed them a food reward. In short order, the apes were all reliably gesturing through the correct hole toward the experimenter. . . .

So how did the animals react when they encountered two familiar experimenters, one who could see them, the other who could not? They entered the lab, but then (measurably) paused. And yet, having apparently noted the novelty of the circumstance, they were then just as likely to gesture to the person who could *not* see them, as to the person who could. This was true in three of the four conditions [with blindfolded experimenters, experimenters with buckets over their heads, and experimenters with hands over their eyes]. In each of these conditions, the chimpanzees displayed no preference for gesturing toward the experimenter who could see them. In contrast, on the easy surrounding trials, the apes gestured through the correct hole (in front of the only experimenter present) 98 per cent of the time. Thus, despite their general interest and motivation, when it came to the seeing/not seeing conditions, the animals appeared oblivious to the psychological distinction between the two experimenters.

There was, however, one exception. Unlike the blindfolds, buckets, and hands-over-the-eyes trials, on back-versus-front trials (in which one person faced toward the ape and the other faced away) the animals gestured to the person facing forward from their very first trial forward. Here, then, the animals seemed to have the right idea: 'Gesture to the person who can see.' But why the discrepancy? Why should the apes perform well on a condition in which one of the experimenters was facing them and the other facing away, but then not on any of the other conditions? In defense of the high-level account, it could be argued that the back/front condition was simply the easiest situation in which to recognize the difference between seeing and not seeing. And, despite the fact that the animals had measurably paused before making their choices in the other conditions as well, and despite the fact that we had observed them adopt these other postures in their play, the idea that back/front was simply a more natural distinction felt appealing.

However, despite the seeming clarity of our intuitions, there was another, more mundane potential explanation of these results. It was possible that on the back/front trials the apes were merely doing what we had, in effect, trained them to do—enter the test lab, look for someone who happened to be facing forward, and then gesture in front of him or her. Rather than reasoning about who could see them, perhaps the apes were simply executing a procedural rule that we had inadvertently taught them. Worse yet, perhaps evolution had simply sculpted them to gesture to the front of others, without any concomitant appreciation that others 'see'.

At this point, several ways of distinguishing between these possibilities occurred to us. If the high-level account were correct (that is, if the back/front condition was simply the most natural case of seeing/not seeing), then the apes ought to perform well on other, equally natural conditions. Here, for example, is another situation that our apes experience on a daily basis. One of our females approaches a group of others who are facing away from her. As she gets closer, one of the other apes turns around and looks over her shoulder toward the approaching animal. Now, although the approaching female notices this behavior, does she understand that the other ape is *psychologically* connected to her in a way that the others are not? The new condition that this consideration inspired ('looking-over-the-shoulder') was of interest in its own right, but we had an even stronger motivation for testing the apes on such a condition. Recall that the low-level account could explain our apes' excellent performance on the back/front condition by posting that they were simply being drawn to the frontal posture of a person. But in this new, looking-over-the-shoulder condition, there was no general frontal posture—just the face of one experimenter and the back of the other one's head. Thus, the low-level account generated the seemingly implausible prediction that the apes would perform well on the back/front condition, but randomly on the looking-over-the-shoulder trials. In contrast, the high-level model predicted the seemingly more plausible outcome in which the apes would gesture to the person who could see them.

To our surprise, however, and in full support of the low-level model, on the looking-over-the shoulder trials the apes did not prefer to gesture to the person who could see them. In direct contrast, they continued to perform without difficulty on the back/front trials. This result made a deep impression on us. No longer was it possible to dismiss our original results by supposing that the animals thought that we were peeking from under the buckets or blindfolds, or between our fingers. No, here we had made 'peeking' clear and explicit, and yet the apes still performed according to the predictions of the low-level model. The experimental dissection of the fronts of the experimenter from their faces (using a posture that our apes must witness every day), sobered us to the possibility that perhaps our animals genuinely might not understand that the experimenters had to *see* their gesture in order to respond to it. More disturbing still, the results seemed to imply that even for the back/front condition our apes might have no idea that the experimenter facing away was 'incorrect'—rather, this was simply a posture with a lower valence. After all, the animals were perfectly willing to choose the person in this posture on fully half of the looking-over-the-shoulder trials.

We had difficulty accepting the implications of these results. We had witnessed our apes using their begging gesture in both testing and non-testing situations on hundreds of occasions, and had always been comfortable in assuming that they conceptualized what they were doing in the same manner that we did. In fact, it was almost impossible not to do so. They would approach us, stick out their hands, and then look up into our eyes. Was it really possible that a behavioral form so instantly recognizable to us could be understood so differently by them?

Thus, despite the fact that the high-level model had done a very poor job at predicting how our apes would react to our tests, we nonetheless remained deeply skeptical of the alternative, low-level model. And so, after further reflection, we decided to examine our animals' reactions to several other conditions, such as one involving screens. In order to go the extra mile, before we began testing them in this condition, we familiarized the apes with the screens by holding the screens in front of our faces and playing 'peek-a-boo' with the animals. We even let the apes play with the screens themselves. And yet despite all of this, when it came to testing, the apes responded in the same manner as they had before; they were just as likely to choose the person who could not see them as the person who could.

SOURCE: From pp. 19–20 and 29–34 in *Folk Physics for Apes*, by Daniel J. Povinelli. Copyright 2000 by Daniel J. Povinelli. Reprinted by permission of Oxford University Press.

rivals from associating with one another, but he was not successful. This time, Nikkie emerged on top, and Yeroen regained some of his former power.

It is difficult to be certain that Yeroen, Luit, and Nikkie consciously plotted their strategies or were aware of the political consequences of their behavior. De Waal believes, however, that it is a reasonable possibility and should seriously be considered. Others have reported similar though less baroque examples of manipulation of coalitions among male chimpanzees.

THEORY OF MIND

The ability to understand the mental states of other individuals is called a theory of mind.

Developmental psychologists have discovered that very young children cannot distinguish between their own knowledge and the knowledge of others. They acquire this ability only as they mature. The capacity to distinguish between the two is called a **theory of mind** and is considered an essential prerequisite for performing complex deceptions, imitating, pretending, and teaching. Recently, primatologists have begun to consider the question of whether or not nonhuman primates can attribute mental states to others.

The question of whether monkeys and apes can attribute mental states to others might seem both impossible to demonstrate conclusively (see Box 9.2) and of no real practical importance. However, a theory of mind is useful to animals living in social groups. For example, deception requires the manipulation of another individual's belief about the world. Consider a low-ranking male who unexpectedly comes upon a desirable food item. He may know from past experience that older and stronger animals routinely take such items from him. It would make sense, then, either to carry the food item away, hide it and return to it later, or to lead unsuspecting group members away from the area. But to execute this deception, the finder must first understand that his knowledge differs from the knowledge of other group members and

Box 9.2

Examining Theory of Mind in Children, Monkeys, and Apes

You might be surprised, skeptical, or extremely doubtful that we would venture to draw conclusions about nonhuman primate ability to distinguish between their own knowledge of the world and others' knowledge of the world. How can we know this for any organism that we can't question directly? Cognitive psychologists have devised a number of extremely clever ways to address these issues that do not require language. Several experiments deal with **attribution,** the capacity to assess the knowledge or mental states of others.

To give you a concrete example, we will briefly outline the classic experiment on attribution. A young child is shown a matchbox and asked what it contains. The child normally answers that it contains matches. The child is shown that the matchbox actually contains something else, say M&M's (or the British equivalent of M&M's called Smarties). Then a newcomer enters the room, and the child is asked what the newcomer will think is in the matchbox. Children below the age of three or so invariably say that the newcomer will think the matchbox contains M&M's. Older children say that the newcomer will think the matchbox contains matches. These results suggest that young children cannot distinguish what they know about the world from what others know about the world, but that they acquire this ability as they mature.

This test cannot be applied directly to monkeys and apes, but other tests can. Cheney and Seyfarth conducted a series of experiments designed to assess attribution in monkeys. One of these experiments compared the behavior of mothers when their offspring were aware of the presence of desirable food items and when they were ignorant of the foods' proximity. In one set of trials mothers and their offspring were seated side by side, and both could see into the test area as apple slices were placed into a food bin. In the other set of trials, only the mother could see the apple slices being placed in the food bin. Then the juvenile was released into the test area. Mothers of ignorant offspring did not behave any differently from mothers of knowledgeable offspring; they did not call more often, orient more toward the food bin, or otherwise seem to communicate their knowledge to their offspring. As a result, knowledgeable offspring found the food items significantly sooner than ignorant offspring did.

These results suggest that the mothers did not differentiate between what they knew and what their infants knew, though other explanations for the mothers' behavior are possible. For example, a mother might be aware of her infant's mental state, but not use this information to alter the infant's behavior. However, this interpretation is weakened by the fact that in a parallel experiment, mothers also failed to warn ignorant offspring about the presence of a frightening and potentially dangerous situation. Certainly, there should be strong selection favoring alerting offspring to the presence of danger.

Woodruff and Premack, from the University of Pennsylvania, conducted an ingenious experiment to assess the chimpanzee's ability to attribute mental states. Sarah, a chimpanzee involved in research on language and cognition, was shown videotapes of an actor in a cage who was faced with a variety of dilemmas. In one case the actor could not reach a bunch of fruit hanging from the top of the cage. In another he could not reach fruit just beyond the bars of the cage. Then Sarah was given a series of photographs. One of the photographs depicted the solution to the problem (such as standing on a chair), and the others showed irrelevant actions (such as reaching sideways with a stick). Sarah routinely chose the photograph that represented the appropriate solution to the problem. Premack argued that Sarah's choice of the correct solution means that she may have understood the actor's intention and desire to get the fruit. To do this, she may have attributed a state of mind to the actor. On the other hand, as Premack acknowledged, it may be that Sarah understood the problem and knew how to solve it, but did not actually attribute a state of mind to the actor.

Sarah could not solve all the attribution problems put to her, but young human children cannot solve all the attribution problems that Sarah can solve. These data do not prove that apes have a theory of mind, or that monkeys do not. But they do provide suggestive evidence regarding this question, and they do indicate that such questions are open to objective, scientific investigation.

then come up with a way to take advantage of this discrepancy effectively. A similar argument can be made for teaching. Female chimpanzees have been seen demonstrating the fundamental elements of nut cracking to their offspring. In teaching, the instructor has to realize the limits of the pupil's knowledge first. In turn, the pupil has to grasp the intent of the instructor as well as the objective of the behavior. Both imitation and teaching play an important role in the transmission of complex behavior patterns in humans and underlie the human capacity for culture.

THE GREAT APE PROBLEM

A considerable body of evidence suggests that great apes have greater cognitive sophistication than monkeys. In particular, chimpanzees seem to have a greater knowledge of the minds of others than monkeys do.

Apes seem to have more knowledge of what others are thinking than monkeys do (see Box 9.2). This capacity may permit apes to manipulate and exploit others, and it may explain why chimpanzees figure so prominently in Byrne and Whiten's compendium of deception (see Box 9.1) and de Waal's accounts of political intrigue. Knowledge of others' state of mind may underlie the capacity for empathy and may enable chimpanzees to console others after conflicts have ended. There is also experimental evidence that chimpanzees are better able to view tasks from others' perspective than monkeys are. Chimpanzees also provide the best evidence of teaching and imitation and are the most proficient tool users (Figure 9.12).

FIGURE 9.12

Chimpanzees make and use tools in the wild. Here a female carefully inserts a twig into a hole in a termite mound. (Photograph by William McGrew.)

The social intelligence hypothesis does not predict that apes will be more intelligent than monkeys.

Richard Byrne, one of the original architects of the social intelligence hypothesis, points out that great apes do not face greater social challenges than other monkeys do, but apes are nonetheless more intelligent than monkeys. The social groups of great apes are no larger and no more complex than the social groups of many monkeys. Orangutans, in fact, are largely solitary. Thus, it seems unlikely that social pressures alone are responsible for cognitive differences between monkeys and apes.

Some researchers believe that the cognitive abilities of great apes evolved in response to ecological challenges, not social challenges.

Byrne suggests that ancestral populations of great apes were once subject to strong selective pressures for more efficient feeding. Great apes have a particular need for efficient feeding techniques because they are large-bodied creatures who move relatively slowly and inefficiently and have no specialized digestive anatomy or cheek pouches to store food. The adaptive solution to this problem was the development of an ability to plan a course of action. This in turn required the ability to represent actions mentally. Byrne suggests that great apes can represent abstract problems in their minds, enabling them to simulate alternative actions and to compute potential outcomes. The ability to visualize and plan is reflected in the development of technically demanding foraging techniques, which only apes can master. Although these abilities were initially favored because they enhanced foraging efficiency, they have come to play an important role in the social lives of great apes as well.

Great apes make use of more complex foraging techniques than other primates do.

Byrne points out that great apes make use of more complicated foraging techniques than other primates do, enabling them to feed on some foods that other primates cannot process. For example, he notes that all plant foods that mountain gorillas rely upon are well defended by spines, hard shells, hooks, and stingers. To process these kinds of foods, gorillas must perform a different sequence of steps, each structured in a particular way. Many of the foods that orangutans feed upon are also difficult to process.

Great apes also use tools to obtain access to certain foods that are not otherwise available to them. Chimpanzees poke twigs into holes of termite mounds and anthills, use leaves as sponges to mop up water from deep holes, and employ stones as hammers to break open hard-shelled nuts (see Chapter 18). Recently, Carel van Schaik of Duke University and his colleagues observed Sumatran orangutans using sticks to probe for insects and to pry seeds out of the husks of fruit. Although bonobos and gorillas have not been observed to use tools in the wild, all apes are adept tool users in captive settings.

A number of animals use tools in the wild, but chimpanzee tool use suggests that more complex cognitive processes are involved when apes use them: two or more tools may be utilized in sequence to perform a single task, one tool may be employed to make another tool, tools may be chosen before they are actually needed, tools may be modified before they are used, and tools may be manipulated in novel ways to fit new circumstances.

Although the special cognitive abilities of great apes may have evolved to solve ecological challenges, these abilities have come to play an important role in the social domain as well. The ability to represent abstract problems mentally, to simulate alternative courses of action, and to compute potential outcomes enables great apes to become adept at social maneuvering. Apes are able to understand the minds, reactions, and likely behavior of other group members; to evaluate alternate tactics; and to manipulate and deceive others effectively.

Some researchers believe that the magnitude of the cognitive gap between monkeys and apes has been exaggerated, and therefore deny that great apes pose a challenge to the social intelligence hypothesis.

Michael Tomasello and Josep Call, who recently published an encyclopedic review of studies of primate cognition, believe that great apes may not actually be smarter than monkeys. They point out that direct comparisons of monkeys' and apes' cognitive skills are scarce because relatively few cognitive experiments have been conducted on both. In the few cases in which direct comparisons can be drawn, apes don't always perform better than monkeys. Moreover, researchers don't always know what cognitive skills underlie observed behaviors, making it impossible to attribute sophisticated cognitive abilities to apes that they may not actually possess. Finally, many examples of apes' superior cognitive abilities come from observing apes who have had extensive contact with humans and are not seen in apes who have been raised in more natural contexts.

The Value of Studying Primate Behavior

As we come to the end of Part Two, it may be useful to remind you of the reasons why information about primate behavior and ecology plays an integral role in the story of human evolution. First, humans are primates, and the first members of the human species were probably more similar to living nonhuman primates than to any other animals on Earth. Thus, by studying living primates we can learn something about the lives of our ancestors. Second, humans are closely related to primates and similar to them in many ways. If we understand how evolution has shaped the behavior of animals so much like ourselves, we may have greater insights about the way evolution has shaped our own behavior and the behavior of our ancestors. Both of these kinds of reasoning will be apparent in Part Three, which covers the history of our own lineage.

Further Reading

Byrne, R., and A. Whiten, eds. 1988. *Machiavellian Intelligence.* Oxford University Press, Oxford.

Byrne, R. W. 1995. *The Thinking Ape.* Oxford University Press, Oxford.

Cheney, D. L., and R. M. Seyfarth. 1990. *How Monkeys See the World.* University of Chicago Press, Chicago.

Russon, A. E., K. A. Bard, and S. T. Parker. 1996. *Reaching into Thought: The Minds of the Great Apes.* Cambridge University Press, Cambridge.

Whiten, A., and R. W. Byrne. 1997. *Machiavellian Intelligence II.* Cambridge University Press, Cambridge.

Study Questions

1. Diet, extractive foraging, and social challenges have all been suggested as influences on the evolution of cognitive abilities among nonhuman primates. What is the rationale for each of these models?
2. Robin Dunbar's analysis comparing primate group size and the relative size of the primate neocortex is based largely on the correlation between these two variables. What shortcomings are inherent in correlational analyses like this one?
3. Compare and contrast the forms of deception displayed by primates and by other animals.
4. What is meant by a theory of mind? Why is this topic important in primate studies?
5. Describe the patterning of alliances among primates. Explain why alliances are such complicated interactions for the participants.
6. What are the advantages and disadvantages of using experimental methods, like those developed by Cheney and Seyfarth, in studying primate social behavior? What advantages do such methods have over naturalistic observations, and vice versa?
7. What evidence do we have that monkeys have a concept of kinship?
8. The social intelligence hypothesis runs into trouble when we come to the great apes. What problem do these species create? What are the possible explanations for the cognitive abilities of the apes?

9. Tomasello and Call believe that some researchers attribute more sophisticated cognitive abilities to primates than they really have, and label such explanations "cognitively generous." Consider the examples of deception in Box 9.1. Provide a "cognitively generous" and "cognitively stingy" explanation for each of these examples. How would you design a study to distinguish between these kinds of explanations for one of these examples?

10. Many species of animals demonstrate very impressive feats of memory, navigation, and technology. Thus, some birds can remember hundreds of different locations where they have stored seeds, paper wasps make intricate nests, and tiny monarch butterflies migrate thousands of miles every year. Why do we think primates' cognitive abilities are more impressive than those of other animals? What features distinguish primates' cognitive abilities from those of other animals?

CHAPTER 18

Evolution and Human Behavior

- Why Evolution Is Relevant to Human Behavior
- Evolutionary Psychology
  - The Logic of Evolutionary Psychology
  - Reasoning about Reciprocity
- Evolutionary Psychology and Human Universals
  - Inbreeding Avoidance
- Evolution and Human Culture
  - Culture Is a Derived Trait in Humans
  - Culture Is an Adaptation
- Human Behavioral Ecology

Why Evolution Is Relevant to Human Behavior

The application of evolutionary principles to understanding human behavior is controversial.

The theory of evolution is at the core of our understanding of the natural world. There is no doubt that every aspect of all living creatures is the product of evolution. By studying how natural selection, recombination, mutation, drift, and other evolutionary processes interact to produce evolutionary change, we come to understand why organisms are the way they are. Of course, our understanding of evolution is far from perfect, and there are other disciplines, most notably chemistry and physics, that contribute greatly to our understanding of life. Nonetheless, evolutionary theory is an essential part of biology and anthropology.

So far, the way we have applied evolutionary theory in this book is not controversial. We are principally interested in the evolutionary history of our own species, *Homo sapiens*, but we began by using evolutionary theory to understand the behavior of our closest relatives, the nonhuman primates. Twenty years ago, when evolutionary theory was new to primatology, this approach generated some controversy, but now most primatologists are committed to evolutionary explanations of behavior. In the section on hominid evolution, we used evolutionary theory to develop models of the patterns of behavior that might have characterized early hominids. While some researchers might debate the fine points of this analysis, the use of evolutionary theory in this context creates little controversy. Perhaps this is because the early hominids were simply "bipedal apes," with brains the size of modern chimpanzees. Not many people object to evolutionary analyses of human physiology like those presented in the last chapter. While scientists argue the merits of particular evolutionary explanations of senescence, few doubt that human physiology has been shaped by natural selection, and that we may gain important insights about how our bodies work if we understand these processes more fully.

It is mainly when we enter the domain of contemporary human behavior that evolutionary analyses provoke intense controversy. Most social scientists acknowledge that evolution has shaped our bodies, our minds, and our behavior to some extent. However, many social scientists are critical of the application of evolutionary theory to contemporary human behavior because they think this implies that contemporary human behavior is determined genetically. Genetic determinism of behavior seems inconsistent with the fact that so much of human behavior is acquired through learning, and that so much of our behavior and beliefs is strongly influenced by our culture and environment.

All phenotypic traits, including behavioral ones, result from interactions between genes and the environment.

The controversy arises, in part, because many people view learning and genetic transmission as mutually exclusive alternatives: behaviors are either genetic and immutable, or learned and controlled entirely by environmental contingencies. This assumption lies at the heart of the "nature-nurture question," a debate that has raged in the social sciences for many years. Despite its persistence, the debate is based on a fundamental misunderstanding of how the natural world works.

The nature-nurture debate assumes there is a clear distinction between the effects of genes (nature) and the effects of the environment (nurture). People often think that genes are like engineering drawings for a finished machine, and people vary simply because their genes carry different specifications. For example, they would think that Bill Lambeer is tall because his genes specified an adult height of 2.11 m (6 ft 11 in.), while Muggsy Bogues is short because his genes specified an adult height of 1.60 m (5 ft 3 in.).

However, this way of thinking about genes is wrong. Genes are not like blueprints that precisely specify the details of the adult phenotype. Every trait results from the *interaction* of some genetic program with the environment. Thus, genes are more like recipes in the hands of a creative cook, sets of instructions for the construction of an organism using materials available in the environment. Genes control the production of a large number of chemicals, mainly enzymes, that interact with each other and the environment to generate the adult phenotype. At each step, this very complex process depends on the nature of environmental conditions. For example, the rate of chemical reactions depends on temperature, the availability of certain chemical substrates, the presence of pathogens, and myriad other environmental factors. As a result, the expression of any genotype always depends on the environment. Thus, we might discover that people are short because of poor nourishment during childhood or because they contracted a disease that limited their growth.

The expression of behavioral traits is usually more sensitive to environmental conditions than the expressions of morphological and physiological traits are. Therefore, we expect language to vary widely among human societies, while we expect finger number and the size of canine teeth to vary little among societies. As we saw in Chapter 3, traits that develop uniformly in a wide range of environments, like finger number, are said to be canalized. Traits that vary in response to environmental cues are said to be plastic. The point to remember is that no trait is purely genetic or purely environmental. Every trait, plastic or canalized, results from the unfolding of a developmental program in a particular environment. Even highly canalized characters can be modified by environmental factors, such as fetal exposure to mutagenic agents.

Natural selection can shape developmental processes so that organisms develop different adaptive behaviors in different environments.

Some people accept that all traits are influenced by a combination of genes and environment, but they reject evolutionary explanations of human behavior because they have fallen prey to a second, more subtle misunderstanding—the belief that evolutionary explanations imply that behavioral differences between individuals must be caused by genetic differences because natural selection cannot create adaptations unless such differences between individuals exist. According to this view, we cannot argue that variation in foraging strategies in different human groups is adaptive without implying that particular foraging strategies have a genetic basis. Since there is considerable evidence that human differences are the product of learning and culture, adaptive explanations must be invalid.

Of course, it *is* true that evolution of adaptations requires genetic variation. However, it does not follow that observed adaptive differences between individuals are the result of genetic differences. In Chapter 3 we saw that male mate-guarding behavior varies adaptively among populations of the soapberry bug in Oklahoma. Where there are many more males than females, males guard females for extended periods, but males do not guard where females are more common. Recall that most of this variation is not genetic; instead, individual males vary their behavior adaptively in response to the local sex ratio. In order for this behavior to evolve, there had to be some genetic variation affecting male mate-guarding behavior—small genetic differences in the male propensity to guard a mated female, and small genetic differences in how this propensity changes with the local sex ratio. If such variation exists, then natural selection can mold the response of males so that it is locally adaptive. However, in any given population, most of the observed behavioral variation is due to the fact that individual males respond adaptively to environmental cues.

Behavior in the soapberry bug is relatively simple. Human learning and decision making are immensely more complex and flexible. People living in varied environments and cultures have profoundly different subsistence strategies, domestic arrangements, child-rearing practices, religious beliefs, political systems, and languages. We know much less about the mechanisms that produce such flexibility in humans than we do about the mechanisms that produce flexibility in mate guarding among soapberry bugs. Nonetheless, such mechanisms must exist, and it is reasonable to assume that they have been shaped by natural selection. This assumption may be wrong, because evolution does not produce adaptation in every case, as we discussed in Chapter 3. However, there is no reason to assume that differences in behavior among humans are the product of genetic differences.

In the remainder of this chapter we survey several ways that evolutionary theory has been used to understand the behavior of modern humans. We begin by discussing attempts to use evolutionary ideas to understand the psychological mechanisms that give rise to human behavior. Next, we consider the relationship between evolutionary theory and human culture, and finally we apply evolutionary theory to understanding the structure of contemporary societies.

Evolutionary Psychology

THE LOGIC OF EVOLUTIONARY PSYCHOLOGY

In recent years, a new field called evolutionary psychology has developed. Its practitioners are committed to using evolutionary theory to understand human psychology and behavior. This research program is based on the following precepts:

- Minds are built up out of a large number of special-purpose mechanisms that solve particular kinds of problems.
- Since brains are expensive to build and to maintain, organisms can't be good at all cognitive tasks. For example, organic computing power that is devoted to reasoning about social problems can't also be used to solve abstract, logical puzzles.
- Natural selection determines the kinds of problems that the brains of particular species are good at solving. To understand the psychology of any species we must know what kinds of problems its members need to solve in nature.
- Complex adaptations evolve relatively slowly.
- People have lived in societies with agriculture, high population density, and stratified social organization for only a few thousand years. They lived in small-scale foraging societies for a much longer stretch of human history.
- To understand human psychology we must determine the kinds of problems humans needed to solve when they lived in small groups as Pleistocene foragers (Figure 18.1).

FIGURE 18.1
Evolutionary psychologists believe the human mind has evolved to solve the adaptive challenges that confront food foragers because this is the subsistence strategy that humans have practiced for most of our evolutionary history.

Even the most flexible strategies are based on special-purpose psychological mechanisms.

Psychologists once thought that people and other animals had a few general-purpose learning mechanisms that allowed them to modify any aspect of their phenotypes adaptively. However, a considerable body of empirical evidence indicates that animals are predisposed to learn some things and not others. Rats learn very quickly to avoid novel foods that make them ill. Moreover, rats' food aversions are based solely on the taste of foods that have made them sick, not the food's size, shape, color, or other attributes. This learning rule seems to be sensible because rats live in a very wide range of environments and usually forage at night. Since their environments vary, they frequently encounter new foods. In order to determine whether a new food is edible, they taste a small amount first and then wait several hours. If it is poisonous, they soon become ill, and they do not eat it again. Rats may pay attention to the taste of foods, instead of other attributes, because it is often too dark to see what they are eating (Figure 18.2).

FIGURE 18.2
Rats initially sample small amounts of unfamiliar foods, and if they become ill soon after eating something, they will not eat it again.

However, there are limits to the flexibility of this learning mechanism. There are certain items that rats will never sample, and in this way their diet is rigidly controlled by genes. Moreover, the learning process is not equally affected by all environmental contingencies. Thus, rats are affected more by the association of novel tastes with gastric distress than they are with other possible associations.

> *It is hard to determine the kind of environment that has shaped the evolution of human reasoning.*

Evolutionary psychologists use the term **environment of evolutionary adaptedness (EEA)** to refer to the social, technological, and ecological conditions under which human mental abilities evolved. Many evolutionary psychologists envision the EEA as being much like the world of contemporary hunter-gatherers.

There are two problems with this notion. First, there is great uncertainty about the rate at which complex adaptations, like psychological mechanisms, can arise. In Chapter 1, we saw that natural selection can sometimes lead to rapid change, but we also saw that vast spans of time can pass without any change at all. We cannot exclude the possibility that new mental mechanisms might have evolved since the origin of agriculture just 10 kya, nor can we reject the possibility that some mental mechanisms might have been shaped by selection long before the origin of humans.

The second problem with the idea that the EEA resembled the environment of modern human foragers is that there is great uncertainty about the ecology and behavior of extinct hominids. As we have seen, some authorities believe that *Homo erectus* and perhaps even earlier *Homo* species were much like contemporary human foragers: They lived in small bands and subsisted by hunting and gathering. They controlled fire, had home bases, and shared food. They could talk to one another and had a body of inherited culture. There are other authorities who think *H. erectus* and even the Neanderthals lived completely unlike modern hunter-gatherers. These researchers think that Neanderthal speech was primitive, and that Neanderthals didn't hunt large game, share food, or have home bases. If early hominids lived lives that resembled those of contemporary foragers, then it is reasonable to think that the human brain has evolved to solve the kinds of problems that also confront modern foragers. For example, since food sharing is important for foragers, it would make sense for humans to have evolved psychological mechanisms to detect and punish freeloaders. On the other hand, if lifeways that characterize contemporary foragers did not emerge until 40 kya, then we might doubt that there was enough time for selection to assemble specialized psychological mechanisms to manage food sharing.

REASONING ABOUT RECIPROCITY

Despite the problems in determining the nature of the EEA, evolutionary psychology is a rapidly growing research area. It has provided novel and productive insights about the nature of human psychology, insights that promise to solve long-standing problems in the social sciences. The following example shows how evolutionary reasoning helps us to understand why people can easily solve social problems but find it much more difficult to solve nearly identical problems in other domains.

> *Human reasoning abilities are content-dependent.*

At one time it was thought that human reasoning was governed by the principles of formal logic. However, research by cognitive psychologists has shown that this assumption is not correct. People reason logically only in limited domains, and human reasoning is **content-dependent,** meaning that the subject matter that people are asked to reason about seems to regulate how they reason.

The following problems, drawn from the work of John Tooby and Leda Cosmides of the University of California, Santa Barbara, will convince you that human reasoning is content-dependent. First, consider the clerical problem in Box 18.1. Study this question for a few minutes and write down the answer. Next, consider the bartender's problem in Box 18.2 and write down your answer on a piece of paper.

If you are like most people, you selected only the "D" card, or perhaps the "D" and "3" cards, for the clerical problem. If you are like most people, you selected the cards labeled "Beer" and "17" for the bartender problem. And, finally, if you are like us and most other people, you got the first problem wrong and the second problem right. The correct answer for the clerical problem is "D" and "5." The correct answer for the bartender's problem is "Beer" and "17."

You will probably be surprised to find out that these two problems are logically identical. In each case you are given the statement

If P, then Q

and in both problems the four cards state

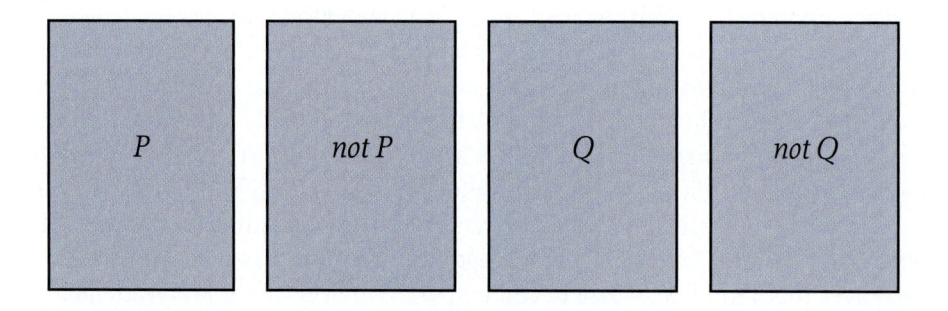

| P | not P | Q | not Q |

Then you are asked which cards you need to turn over to determine whether the statement is true for that card. The laws of logic say that the statement *If P, then Q* can only be falsified by observing *P* and *not Q*. Thus, the only two cards that need to be turned over are *P*, to see that if the other side is *not Q*, and *not Q*, to see if the other side is *P*. Most people can easily see that this reasoning is correct in the case of the bartender problem, but not in the case of the clerical problem.

Some recent data suggest that humans are particularly able to solve problems involving social exchange.

The psychologists who first discovered this effect believed that people were better at solving familiar problems than unfamiliar problems. Since most people have been carded in bars, and few people have done clerical work, the bartender problem is easier to solve. However, additional experiments show that familiarity does not predict which problems people will find easy to solve.

Recently, Cosmides and Tooby provided a different explanation. They contend that people are good at solving problems involving reciprocal altruism and other forms of

Box 18.1
Clerical Problem

Part of your new clerical job at the local high school is to make sure that student documents have been processed correctly. Your job is to make sure that the documents conform to the following rule:

If a person has a 'D' rating, then his documents must be marked code '3'.

You suspect that the secretary whom you replaced did not categorize the students' documents correctly. The cards below have information about the documents of four people who are enrolled at this high school. Each card represents one person. One side of a card tells a person's letter rating and the other side of the card tells that person's number code.

Indicate only those card(s) you definitely need to turn over to see if the documents of any of these people violate this rule.

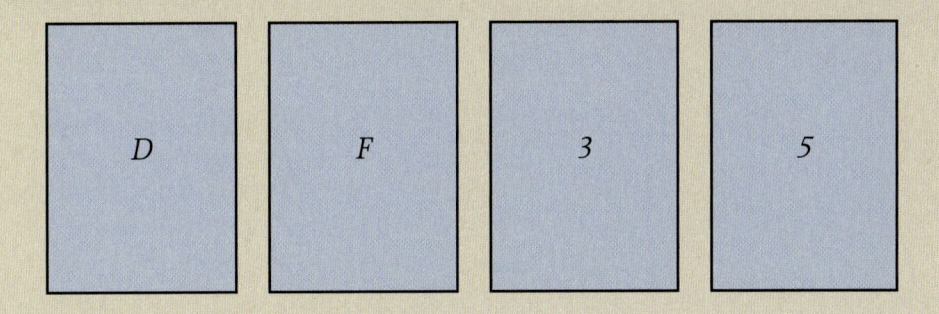

| D | F | 3 | 5 |

From Figure 3.3 in L. Cosmides and J. Tooby, 1992, Cognitive adaptations for social exchange, in *The Adapted Mind, Evolutionary Psychology and the Generation of Culture*, ed. by J. Barkow, L. Cosmides, and J. Tooby. Oxford University Press, New York.

---

social exchange, and this skill was crucial for success in the small groups that have characterized human societies for most of our history. Food sharing, which is an essential part of hunter-gatherer ecology, is a form of reciprocal altruism. The big problem with reciprocal altruism is that it is costly to interact with individuals who do not reciprocate. Thus, human cognition should be tuned to look for cheaters in social exchanges. The bartender problem that you were asked to solve in Box 18.2 has the form "If you take the benefit, then you pay the cost." In this case, beer is the benefit, and being over 21 is the cost. Tooby and Cosmides argue that people are tuned to attend to situations in which people take the benefit without paying the cost. Thus, they look for people who are drinking beer even though they are under 21.

According to Cosmides and Tooby, most people get the logically correct answer on the bartender problem because social reasoning and logic just happen to coincide, not because most people are familiar with being carded in bars. If the logical answer and the social reasoning solution are opposed, what solution would people choose? To answer this question, Cosmides and Tooby also asked students to solve what they called the "switched social contract," which takes the form "If you pay the cost, then you get

Box 18.2
Bartender's Problem I

In their crackdown against drunk drivers, state law enforcement officials are revoking liquor licenses left and right. You are a bouncer in a local bar, and you'll lose your job unless you enforce the following law:

If a person is drinking beer, then he must be over 21 years old.

The cards below have information about four people sitting at a table in your bar. Each card represents one person. One side of a card tells what a person is drinking and the other side of the card tells that person's age.

Indicate only those card(s) you definitely need to turn over to see if any of these people are breaking this law.

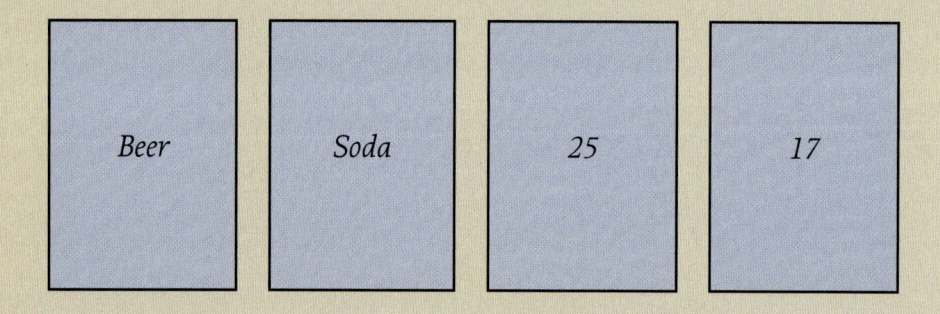

| Beer | Soda | 25 | 17 |

Adapted from Figure 3.3 in L. Cosmides and J. Tooby, 1992, Cognitive adaptations for social exchange, in *The Adapted Mind, Evolutionary Psychology and the Generation of Culture*, ed. by J. Barkow, L. Cosmides, and J. Tooby. Oxford University Press, New York.

---

the benefit." For example, "If a person is over 21, then he may drink beer." This problem still has the logical form *If P, then Q*, but now *P* is "21" and *Q* is "drink beer." Box 18.3 portrays this problem. Again, try to solve this problem yourself.

According to the laws of formal logic, to falsify this statement, you must observe *P* and *not Q*. Thus, in this version of the problem, the logical solution requires you to turn over the "25" card to see if the other side is "not drinking beer," and the "Soda" card to see if the other side is "over 21." However, if you are like most people, you turned over the cards "17" and "Beer." According to Tooby and Cosmides, people get this problem and other problems like this wrong because they have a strong tendency to look for cheaters in social exchange, people who take the benefit (drink beer) without paying the cost (being over 21). People choose the cards that the bartender would need to turn over to detect underage drinkers, not the cards that meet rules of formal logic. (If this is hard to grasp, don't despair. If Cosmides and Tooby are right, then this is simply not a distinction that the human mind has evolved to understand.)

Although Cosmides and Tooby's results are intriguing, some psychologists think that there may be other explanations for them. However, all of the alternate explana-

Box 18.3
Bartender's Problem II

In their crackdown against drunk drivers, state law enforcement officials are revoking liquor licenses left and right. You are a bouncer in a local bar, and you'll lose your job unless you enforce the following law:

If a person is over 21 years old, then he may drink beer.

The cards below have information about four people sitting at a table in your bar. Each card represents one person. One side of a card tells what a person is drinking and the other side of the card tells that person's age.

Indicate only those card(s) you definitely need to turn over to see if any of these people are breaking this law.

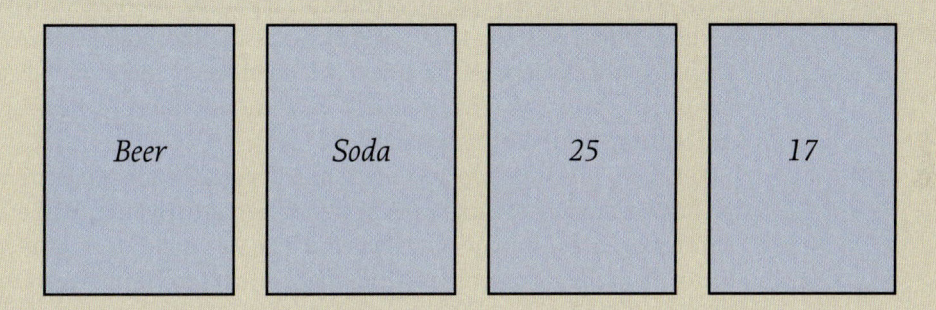

| Beer | Soda | 25 | 17 |

Adapted from Figure 3.3 in L. Cosmides and J. Tooby, 1992, Cognitive adaptations for social exchange, in *The Adapted Mind, Evolutionary Psychology and the Generation of Culture*, ed. by J. Barkow, L. Cosmides, and J. Tooby. Oxford University Press, New York.

---

tions that have been offered presume that people are predisposed to reason about different problems in different ways.

Evolutionary Psychology and Human Universals

Evolved psychological mechanisms cause human societies to share many universal characteristics.

Much of anthropology (and other social sciences) is based on the assumption that human behavior is not effectively constrained by human biology. To be sure, people have to obtain food, shelter, and other resources necessary for their survival and reproduction. But beyond that, human behavior is not constrained by biology.

This assumption is not very plausible from an evolutionary perspective. Evolutionary reasoning suggests that humans should be better at learning some kinds of tasks, making some kinds of decisions, and solving some kinds of problems than others.

From the perspective of evolutionary psychology, it is likely that evolved mechanisms in the human brain channel the evolution of human societies and human culture, making some outcomes much more likely than others. It is not that natural selection has led to genetic differences among different societies. Instead, the mental mechanisms that underly human behavior are held in common among people all over the world; they are universal features of the human species. Tooby and Cosmides liken these universal cognitive features to organs like the liver or the kidneys. Everyone has a liver and two kidneys, and these organs perform the same function from Berkeley to Bora Bora. Similarly, everyone's brain is composed of the same set of **mental organs,** and in every society these mental organs constrain and shape peoples' thoughts, perceptions, and behaviors.

In the remainder of this section, we consider one example of such **human universals,** human inbreeding avoidance.

INBREEDING AVOIDANCE

Incest and its avoidance play a crucial role in many influential theories of human society. Thinkers as diverse as Sigmund Freud, the founder of psychoanalysis, and Claude Levi-Strauss, the father of structuralist anthropology, have asserted that people harbor a deep desire to have sex with members of their immediate family. According to this view, only the existence of culturally imposed rules against incest save society from these destructive passions.

This view is not very plausible from an evolutionary perspective. There are compelling theoretical reasons to expect that natural selection will erect psychological barriers to incest, and good evidence that it has done so in humans and other primates. Both theory and observation suggest that the family is not the focus of desire; it's a tiny island of sexual indifference.

The offspring of genetically related parents have lower fitness than the offspring of unrelated parents do.

Geneticists refer to matings between relatives as **inbred matings,** and contrast them with **outbred matings** between unrelated individuals. The offspring of inbred matings are much more likely to be homozygous for deleterious recessive alleles than are offspring of outbred matings. As a consequence, they are less robust and have higher mortality than the offspring of outbred matings. In Chapter 16 we discussed a number of genetic diseases like PKU, Tay-Sachs, and cystic fibrosis that are caused by a recessive gene. People who are heterozygous for such deleterious recessive alleles are completely normal, while people who are homozygous suffer severe, often fatal consequences. Recall that such alleles occur at low frequencies in most human populations. However, there are many loci in the human genome. Thus, even if the frequency of deleterious recessives at each locus is very small, there is a good chance that everybody has at least one lethal recessive gene somewhere in their genome. Geneticists have estimated that each person carries the equivalent of two to five lethal recessives. Mating with close relatives is deleterious because it greatly increases the chance that both partners will carry a deleterious recessive at the same locus. Theory and experimental data indicate that inbreeding leads to substantial decreases in fitness (Box 18.4). This, in turn, suggests that natural selection should favor behavioral adaptations that reduce the chance that inbreeding occurs.

Matings between close relatives are very rare among nonhuman primates.

Remember from Chapter 7 that in all species of nonhuman primates, members of one or both sexes leave their natal groups near the time of puberty. It is very likely that dispersal is an adaptation to prevent inbreeding. In principle, primates could remain in natal groups and simply avoid mating with close kin. In practice, this is problematic because males and females are likely to know their maternal kin but unlikely to know their paternal kin. To avoid the deleterious consequences of inbreeding, members of one sex must leave the group and seek mating opportunities elsewhere.

Natural selection has provided at least some primates with another form of protection against inbreeding—a strong inhibition against mating with close kin. In a large captive colony of Barbary macaques, German primatologists Jutta Kuester, Andreas Paul, and their colleagues have demonstrated that maternal kin ties have a strong influence on sexual attraction (Figure 18.3). Maternal kin who had lived together in the group since birth very rarely mated, and very few infants were conceived in such unions. However, there was no aversion to mating among maternal relatives who had been separated at some point in their lives, among paternal kin, or among members of different matrilineages. Interestingly, males did not mate with females they had cared for as infants. At Gombe Stream National Park, Tanzania, where chimpanzees have been studied for more than three decades by Jane Goodall and her coworkers, many adult females have contact with their sons or brothers, and some young females in the study group may have been sired by older male residents. Thus, when these females come into estrus, they have opportunities to mate with close kin. In fact, they rarely do. Although mothers have relaxed and affectionate interactions with their sons, matings among them are very rare. There are more brother-sister matings than mother-son matings, but these too are uncommon and generally unwelcome to females. Female chimpanzees seem to have a general aversion to mating with males much older than they are, and males seem to be generally uninterested in females much younger than themselves. These mechanisms may protect females from mating with their fathers, and vice versa.

FIGURE 18.3
Barbary macaques avoid mating with close kin, and this aversion is apparently based on familiarity during early life.

Humans rarely mate with close relatives.

During the first half of the 20th century, cultural anthropologists fanned out across the world to study the lives of exotic peoples. Their hard and sometimes dangerous work has given us an enormous trove of information about the spectacular variety of human lifeways. They found that domestic arrangements vary greatly across cultures: some groups are polygynous, some monogamous, and a few polyandrous. Some people reckon descent through the female line and are subject to the authority of their mother's brother. In some societies, married couples live with the husband's kin, in others they live with the wife's kin, and in some they set up their own households. Some people must marry their mother's brothers' children, while others are not allowed to do so.

Box 18.4
Why Inbred Matings Are Bad News

The following simple example illustrates why inbreeding increases the chance that offspring will be homozygous for a deleterious recessive gene. Suppose Cleo has exactly one lethal recessive somewhere in her genome. If she mates with a nonrelative, Mark, the chance that he carries the same deleterious recessive is simply equal to the frequency of heterozygotes carrying that gene in the population. (We don't need to consider the frequency of individuals who are homozygous for the recessive allele, because it is lethal in homozygotes.) Let p be the frequency of deleterious recessives in the population. Using the Hardy-Weinberg law, we know that the frequency of heterozygotes is $2p(1-p)$. The frequency of lethal recessives is typically about 0.001. Thus, there is only about one chance in 500 that Mark and Cleo carry the same deleterious recessive. If Cleo mates with her brother, Ptolemy, the story is quite different. Remember from Chapter 8 that r, the coefficient of relationship, gives the probability that two individuals will inherit the same gene through common descent. For full siblings $r = 0.5$. Thus, if Cleo mates with her brother, there is one chance in two that he will carry the same lethal recessive that she does, a value that is about 100 times greater than for an outbred mating. Since Cleo and Ptolemy are both heterozygotes for the lethal recessive, Mendel's laws say that 25% of their offspring will be homozygotes and therefore will die. Thus on average, Cleo will experience a 12.5% decrease in fitness by mating with her brother Ptolemy compared with an outbred mating with Mark.

This simple example actually understates the cost of inbreeding because people often carry more than one deleterious recessive. Oxford University biologist Robert May derived the estimates in Figure 18.4 using a more realistic model. As you can see, inbreeding is quite bad. Matings between siblings or between parents and their offspring ($r = 0.5$) produce about 40% fewer offspring than do outbred matings. The effects for other inbred matings are smaller but still serious.

Empirical studies strongly support these predictions. Data from 38 species of captive mammals indicate that the offspring of matings among siblings and among parents and their offspring are 33% less likely to survive to adulthood than the offspring of outbred matings. Studies in wild populations, though less conclusive, tell the same story. Several studies suggest that humans are affected by inbreeding in the same way. For example, the 161 children of father-daughter or brother-sister matings studied by geneticist Eva Seemanová were twice as likely to die during their first year as their maternal half-siblings, and 10 times as likely to suffer serious congenital defects. Other evidence comes from studies of Moroccan Jews living in Israel who do not consider marriages between men and their nieces to be incestuous. A study of 131 children from such marriages indicates that they suffer a 20% reduction in fitness compared with a control group from the same population.

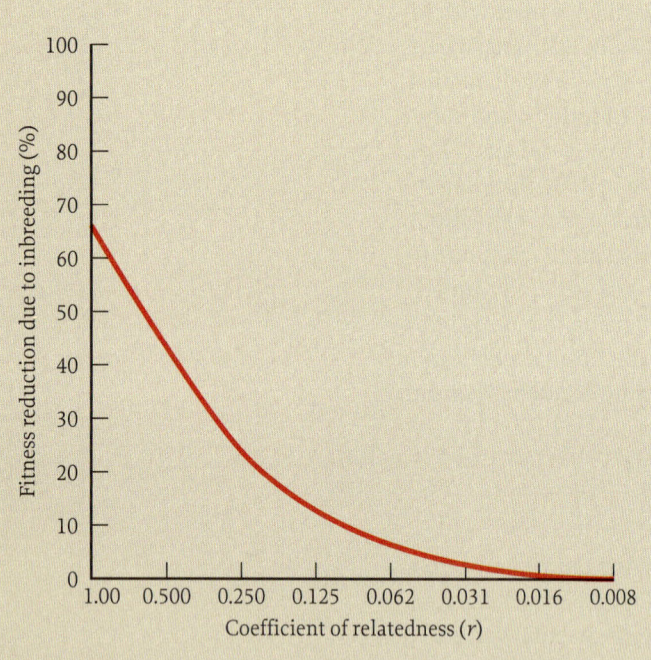

FIGURE 18.4
Population genetics theory predicts that close inbreeding can lead to serious reductions in fitness. The vertical axis plots the percentage of reduction in fitness due to inbreeding, and the horizontal axis plots the coefficient of relatedness among spouses. The curve is based on the conservative assumption that people carry the equivalent of 2.2 lethal recessives.

In all of this almost endless variety of domestic arrangements, there is not a single ethnographically documented case of a society in which brothers and sisters regularly marry, or one in which parents regularly mate with their own children. The only known case of regular brother-sister mating comes from census data collected by Roman governors of Egypt from A.D. 20 to 258. From the 172 census returns that have survived, it is possible to reconstruct the composition of 113 marriages: 12 were between full siblings and eight between half-siblings. These marriages seem to have been both legal and socially approved, as both prenuptial agreements and wedding invitations survive.

The pattern for more distant kin is much more variable. Some societies permit both sex and marriage with nieces and nephews or between first cousins, while other societies prohibit sex and marriage among even distant relatives. Moreover, the pattern of incest prohibitions in many societies does not conform to genetic categories. For example, even distant kin on the father's side may be taboo in a given society, while maternal cousins may be the most desirable marriage partners in the same society. Sometimes the rules about who can have sex are different from the rules governing who can marry.

Adults are not sexually attracted to the people with whom they grew up.

Until recently, most cultural anthropologists thought that cultural rules against incest were the only thing preventing people from doing what they might do otherwise. However, this seems unlikely. The fact that inbreeding avoidance is universal among primates indicates that our human ancestors also had psychological mechanisms preventing them from mating with close kin. These psychological mechanisms could disappear during human evolution only if they were selected against. However, mating with close relatives is highly deleterious in humans and in other primates. Thus, both theory and data predict that modern humans should have psychological mechanisms that reduce the chance of close inbreeding, at least in the small-scale societies in which human psychology was shaped.

There is evidence that such psychological mechanisms exist. In the late 19th century the Finnish sociologist Edward Westermark speculated that childhood propinquity stifles desire. By this he meant that people who live in intimate association as small children do not find each other sexually attractive as adults. Three natural experiments indicate that this mechanism causes inbreeding avoidance in humans.

- *Taiwanese minor marriage.* Until recently, an unusual form of marriage was widespread in China. In **minor marriages,** children were betrothed, and the prospective bride was adopted into the family of her future husband during infancy. There, the betrothed couple grew up together like brother and sister. According to Taiwanese informants interviewed by Stanford University anthropologist Arthur Wolf, the partners in minor marriages found each other sexually unexciting. Sexual disinterest was so great that fathers-in-law sometimes had to beat the newlyweds to convince them to consummate their marriage. Wolf's data indicate that minor marriages produced about 30% fewer children than did other arranged marriages (Figure 18.5a), and were much more likely to end in separation or divorce (Figure 18.5b). Infidelity was also more common in minor marriages. When modernization reduced parental authority, many young men and

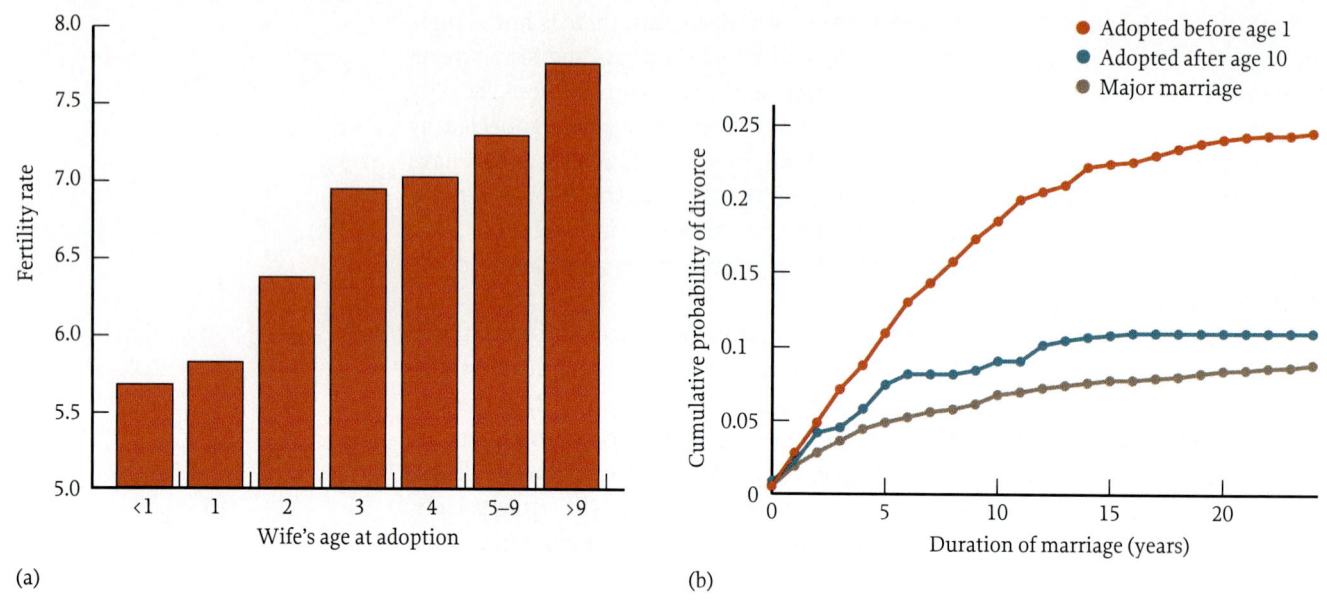

FIGURE 18.5

In minor marriages, the age of the wife when she arrives into her future husband's household (age of adoption) affects both fertility and the likelihood of divorce. (a) The fertility of women adopted at young ages is depressed. (b) The younger women are when they arrive in their husband's household, the less likely it is that their marriages will survive.

women who were betrothed in minor marriages broke their engagements and married others.

- *Lebanese parallel-cousin marriage.* Anthropologist Justine McCabe studied life in a small town in southern Lebanon during the 1970s. The people in this village did not consider marriage between cousins to be incestuous. In fact, they thought that the ideal marriage was between the offspring of brothers, called **patrilateral parallel cousins** by anthropologists (Figure 18.6). Social life in the traditional Arab village studied by McCabe was organized around groups of relatives related through the male line. Brothers often lived together, and even if they did not live in a communal household, they were often best friends and confidants, and their families maintained daily contact with each other. Thus, patrilateral parallel

FIGURE 18.6

Patrilateral parallel cousins are the offspring of siblings of the same sex. In this hypothetical lineage, there are three children, two males (triangles) and a female (circle). The brothers marry and produce children, and their children are patrilateral parallel cousins. Their sister also marries and has children, but her children are not patrilateral parallel cousins of her brothers' children.

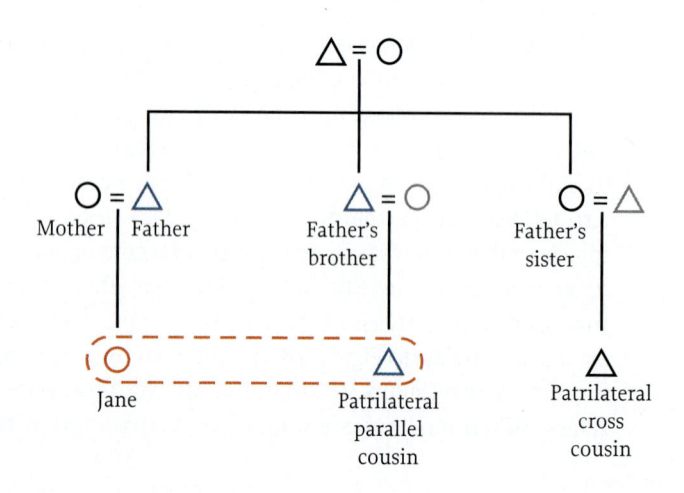

cousins usually grew up much as brothers and sisters do. McCabe compared a large sample of marriages between different kinds of cousins (offspring of sisters, offspring of a brother and a sister, and offspring of brothers). Like Chinese minor marriages, the marriages between patrilateral parallel cousins produced about 20% fewer offspring than did marriages between other kinds of cousins, and were also more likely to end in divorce than were other kinds of marriages. McCabe discovered that people in the town thought that patrilateral parallel cousins who married were less sexually attracted to each other than were other married couples.

- *Kibbutz age-mates.* Before World War II, many Jewish immigrants to Israel organized themselves into utopian communities called **kibbutzim** (plural of **kibbutz**). In these communities, children were raised in communal nurseries, and they lived intimately with a small group of unrelated age-mates from infancy to adulthood. The ideology of the kibbutzim did not discourage sexual experimentation or marriage by children in such peer groups, but neither behavior occurred. Israeli sociologist Joseph Sepher, himself a kibbutznik, collected data on 2769 marriages in 211 kibbutzim. Only 14 of them were between members of the same peer group, and in all of these cases, one partner joined the peer group after the age of six. From data collected in his own kibbutz, Sepher found no instance of premarital sex among members of the same peer group.

Taken together, all of these data suggest that close inbreeding is almost unknown in human societies because people are not sexually attracted to close relatives. These data suggest that people have an innate psychological mechanism causing them to find their childhood companions sexually unattractive. Since brothers and sisters are raised together almost everywhere, people look beyond their parents' household for mates.

The relationship between an aversion to inbreeding and culturally transmitted rules against incest is unclear.

As we discussed earlier, virtually every society has rules that specify acceptable sexual partners for every person. Having sex within the family is incestuous in almost every society, but the rules for more distant kin are quite variable. Taboos sometimes prohibit sex with some very distant relatives, while allowing it with other relatives who are much more closely related.

The relationship between the propinquity effect and the cultural evolution of incest rules is controversial. Westermark thought that this mechanism made incest rare and unattractive. Incest generated feelings of disgust, which in turn caused people to adopt cultural rules against it. The obvious problem with this argument is that the same mechanism ought to lead to an aversion to minor marriage in China and to patrilateral-parallel-cousin marriage in the Middle East. Yet both these practices have persisted for a long time. Other authorities think that people recognize the deleterious effects of inbreeding and consciously adopt rules against it. This view is supported by the observation that people in many societies believe that incest leads to sickness and deformity. However, this explanation does not explain why in some societies marriages between virtually unrelated people are often prohibited while marriages among first cousins are allowed. Finally, many anthropologists think that incest rules are really about marriage and alliances between families, and not about sex at all.

Evolution and Human Culture

For many anthropologists, culture is what makes us human. Each of us is immersed in a cultural milieu that influences the way we see the world, that shapes our beliefs about right and wrong, and that endows us with the knowledge and technical skills to get along in our own environment. Despite the central importance of culture in anthropology, there is little consensus about how or why culture arose in the evolution of the human lineage. In what follows, we present a view of the evolution of human culture developed by one of us (R. B.). Although we believe strongly in this approach to understanding the evolution of culture, there is not a broad consensus among anthropologists that this, or any other particular view of the origins of culture, is correct.

Before we begin, we must provide some definitions of terms. We define **culture** as information acquired by individuals through some form of social learning. Cultural variation is the product of differences among individuals that exist because they have acquired different behaviors as a result of some form of social learning. Culture sometimes has quite different properties from other forms of environmental variation. If people acquire behavior from others through teaching or imitation, then culturally transmitted adaptations can gradually accumulate over many generations (Figures 18.7 and 18.8). For example, one man may learn to fletch his arrows from his father, and his son may learn to dip the arrow point in poison from a neighbor; his grandsons may both fletch their arrows and dip them in poison. To understand culture, we need to take its cumulative nature into account.

Culture Is a Derived Trait in Humans

There has been much debate about whether other animals have culture.

Anthropologists and psychologists have long debated whether other animals have culture. Some authors deny culture to other animals on the grounds that traditions observed in other animals lack the essential features of human culture, such as sym-

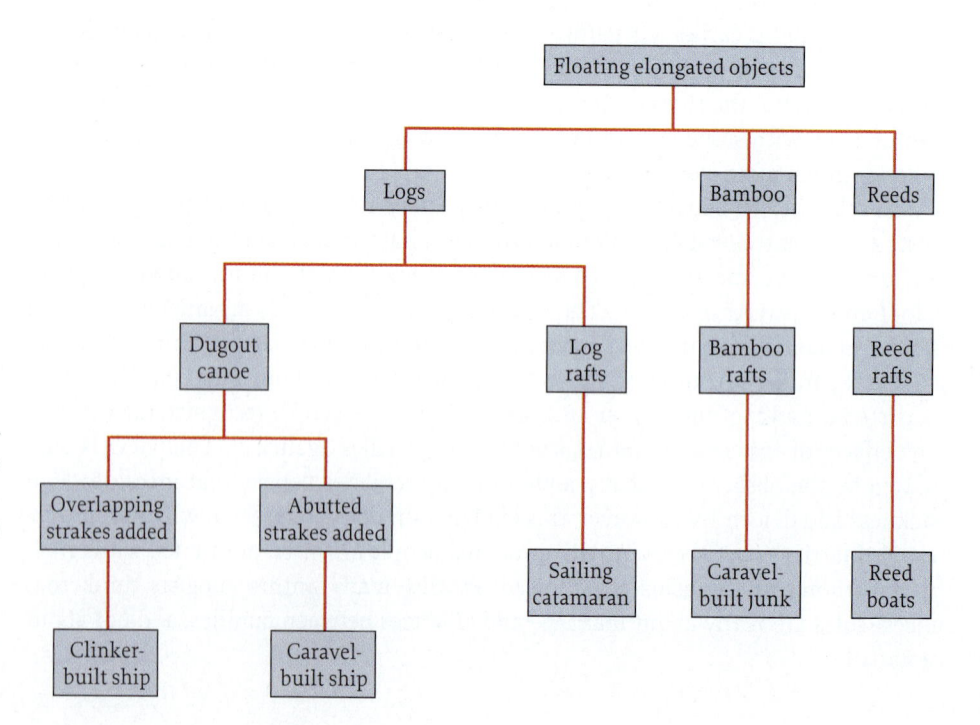

FIGURE 18.7
We can trace the development of certain technological innovations through time. In China, the first boats were elongated structures that floated on the water. Some were made of logs, others of bamboo or reeds. Floating logs were transformed into canoes with the addition of a keel. Other types of ships, including Chinese junks, have a square hull and no keel.

EVOLUTION AND HUMAN CULTURE

(a) (b)

FIGURE 18.8

(a) Bamboo rafts like the one shown here may have been the precursors of (b) the great Chinese junks.

bolic coding. Others argue that those who deny culture to nonhuman animals are applying a double standard—if the kind of behavioral variation observed among some primate populations were observed among human populations, they argue, anthropologists would regard it as cultural.

Such debates make little sense from an evolutionary perspective. The psychological capacities that give rise to human culture are likely to have homologies in the brains of other primates, and the function of cultural transmission in humans could well be related to its function in other species. The study of the evolution of human culture must be based on categories that allow human cultural behavior to be compared with potentially homologous, functionally related behavior of other organisms. At the same time, such categories should be able to distinguish between human behavior and the behavior of other organisms because it is quite plausible that human culture is different in important ways from related behavior in other species.

Culture is common among other animals, but cumulative cultural change is rare.

There are many examples of cultural variation in nature, particularly among primates (Box 18.5). There is little evidence, however, of cumulative cultural evolution in other species. In most cases, social learning only leads to the spread of behaviors that individuals could have learned on their own.

Scientists who study cultural behavior in other organisms distinguish two different classes of mechanisms that can account for cultural differences between populations:

1. **Social facilitation** occurs when the activity of older animals indirectly increases the chance that younger animals will learn the behavior on their own. Young individuals do not acquire the behavior by observing older individuals. Social facilitation would account for the persistence of tool use in the following scenario. In

Cultural Diversity and Human Universals

Many anthropologists, probably most of them, are skeptical of statements that generalize about what all peoples do. But are there not generalizations of that sort that really do hold for the wide array of human populations? There are—and not enough has been said about them. This skepticism and neglect of human universals is the entrenched legacy of an "era of particularism" in which the observation that something *doesn't* occur among the Bongo Bongo counted as a major contribution to anthropology.... The truth of the matter is, however, that anthropologists probably always take for granted an indefinite collection of traits that add up to a very complex view of human nature. Let me give some examples.

In a course that I teach on the peoples and cultures of Southeast Asia I have often illustrated the cultural elaboration of rank that is found in many Southeast Asian societies—and certainly among the Brunei Malays with whom I did my doctoral research—with the following anecdote. In the course of my research I was once seated with two young men on a wooden bench at the front of the house that my wife and I rented in a ward of the Brunei capital. A third young man was seated just a few feet away on the rung of a ladder but at the same height as the rest of us. There was no one else around. Tiring of sitting on the bench, I slipped down from it to sit on the walkway. I was followed almost instantly by all three of the young men. Just as quickly I realized that they had done it not because they too were uncomfortable on the bench (I had been there longer than they) but because in the Brunei scheme of things it is not polite to sit higher than another person, unless you considerably outrank that other person. So I protested, urging them to please remain seated on the bench. They said it wouldn't look nice. I said there was no one but us around to notice. One of them closed the matter by noting that people across the river—to which he gestured (it was about a quarter mile away)—just *might* see what was going on. The clear implication was that he and his fellows weren't about to let anyone see them apparently breaking one of the important rules in the etiquette of rank, even though they knew they wouldn't be offending me.

I always told this story to illustrate difference, to show the extremity to which Bruneis concerned themselves with rank, and it always seemed to be a very effective message. As a teacher of anthropology I know very well that cultural differences elicit some sort of inherent interest. Ruth Benedict's *Patterns of Culture* (1934) is an all-time anthropological best-seller, and its essential message is the astonishing variability of human customs. No one teaching anthropology can ignore the way students react to revelations about the amazing ways other peoples act and think. And no one teaching anthropology can fail to sense the wheels turning in students' minds when they use these revelations to rethink the ways people act and think in their own society. Teachers of anthropology not only see this in students, they cultivate it. But are the differences all that should be of concern to anthropology? Does an emphasis on differences present a true image of humanity?

I now realize that the story I have told my students is pervaded with evidence of similarities: above all, the young men were concerned with what other people would think about them; they were also concerned with politeness in particular, rules in general; even their concern with rank was only a matter of difference in degree. I could go on, mentioning their use of language and gestures; the smooth conversational turn taking; the concepts of question, answer, explanation; the use of highness/lowness to symbolize rank; and much more.

At a more subtle level, I believe, some amazing things were happening that I took no note of. Without my explaining things in detail, in my broken Malay, the young men had instantly grasped my point: the setting was informal and I wanted them to treat me as they would treat each other (they would not have moved down or up in unison for each other in those circumstances); furthermore, it was "not my custom" to be offended by people sitting higher than me. I think that my companions sized up these aspects of the immediate situation just as I had.

But they also saw a wider context in which their behavior could be misinterpreted by others, and with what seemed like a few words and a gesture, they explained their position to me and closed the matter. There were more than a few words and gestures: there were tone of voice, facial expressions, body language, and an enormously complex context of past, present, and future. And there were four human minds, each observing, computing, and reacting to the "implicature" ... of the bare words so silently and automatically as to occasion no notice. All this—from the conscious concern with what others would think to the unconscious assessments of implications—formed a plainly human background, from which I in my lectures had pulled out a quantitative difference as the focus of attention.

I use the word "quantitative" because, although it may not be *my* custom to think that the height of one's seat should match one's rank, the idea is not foreign to western culture. There are some wonderful examples of the equation between seating height and rank, or dominance, in Charlie Chaplin's film "The Great Dictator." What distinguishes the Bruneis from us is the greater frequency of day-to-day contexts in which the equation is observed among Bruneis.

Now it might be objected that the Brunei Malays are so westernized that of course they are similar to us in many ways; one needs a pristine, uncontacted people to see the real exceptions. This is an assumption that I would have taken quite seriously at one time, and that was acted upon by my university schoolmate Lyle Steadman. ... Like me, he received his anthropological training in the 1960s and was steeped in cultural determinism. In order to fully explore the consequences of having a nonwestern worldview, he did his work among a New Guinean people, the Hewa, who had had no more than the most fleeting and widely spaced contacts with European patrols. At the time Steadman studied the Hewa they lived in one of the last "restricted" areas of New Guinea. This meant that the area was "uncontrolled," and Europeans, including missionaries, were forbidden to enter it. One of the reasons the Hewa were essentially uncontacted was that they lived so sparsely on the land that from one family's household to another was typically a grueling 2-hour walk over a rugged terrain covered by dense rain forest.

Steadman had to learn the Hewa language in the field, but long before he was conversant in it he discovered—somewhat to his surprise, because it didn't jibe with his assumptions about the influence of differing worldviews—that he and the Hewa "could understand each other well enough to live together." ... As time went by, and he learned more about the ways in which the world is put together differently in Hewa than in English, he was led to observe that the differences were largely superficial: "This fact of experiencing the world in a similar way," in spite of its being carved up differently in different languages, "became increasingly obvious as I acquired greater proficiency in the language." At the deeper level of why language might be used in the first place, at the level of motives, the similarities were just as evident: "Living, travelling, working and hunting with the Hewa, made it clear to me that their basic concerns, the concerns motivating their behaviour, were similar to my own." ...

SOURCE: From pp. 1–3 in D. Brown, 1991, *Human Universals*, McGraw-Hill, New York. Reproduced by permission of The McGraw-Hill Companies.

populations where chimpanzees use tools to crack nuts, young chimpanzees spend a lot of time in proximity to both nuts and hammer stones. Nuts are a greatly desired food, and young chimpanzees find eating nut meats highly reinforcing. They experiment with stone hammers and anvils until they master the skill of opening the nuts. In populations in which chimpanzees do not use stones to open nuts, young chimpanzees never spend enough time in proximity to both nuts and hammer stones to acquire the skill.

2. **Observational learning** occurs when younger animals observe the behavior of older animals and specifically learn how to perform an action by watching others. In this case, the tool tradition is preserved because young chimpanzees actually imitate the behavior of the older ones.

Box 18.5
Examples of Culture in Other Animals

These are just a few of the hundreds of well-documented examples of cultural variation in nonhuman species.

POD DIPPING Marc Hauser of Harvard University observed an old female vervet monkey dip an *Acacia* pod into a pool of liquid that had collected in a cavity in a tree trunk. She soaked it for several minutes and then ate the pod. This behavior had never been seen before, though this group of monkeys had been observed regularly for many years. Within nine days, three other members of the old female's family had dipped pods themselves, and so had one unrelated adult female. Ultimately, seven of the 10 members of the group learned to dip pods.

GROOMING POSTURES Chimpanzees in the Mahale Mountains of Tanzania often adopt a unique grooming posture (Figure 18.9). During grooming sessions, both partners may simultaneously extend one arm over their heads. The two partners clasp hands, and then groom one another's exposed armpits. These grooming handclasps occur often and are performed by all members of the group. Chimpanzees at Gombe, who live less than 100 km (62.5 miles) away in a similar type of habitat, groom often but never perform this behavior.

TERMITING TOOLS Chimpanzees at several sites use probes made from plant materials to fish for termites (Figure 18.10). At Gombe in Tanzania and Mt. Assirik in Senegal, chimpanzees sometimes use woody vegetation to make termiting tools. At Mt. Assirik, woody twigs and vines are peeled to remove the bark, the bark is discarded, and the peeled twig is used to fish for termites. At Gombe, twigs and vines are sometimes peeled, but when they are peeled, the bark is used to fish for termites. In this case, both populations have access to similar kinds of vegetation for termiting tools, and chimpanzees in both populations know how to peel the twigs. But in one population, the bark is used as the tool, and in the other population the bark is discarded.

FIGURE 18.9
Chimpanzees in the Mahale Mountains often hold their hands above their head and clasp their partner's hand as they groom. This grooming posture has never been seen at Gombe, just 100 km (62.5 miles) away. (Photograph courtesy of William C. McGrew.)

FIGURE 18.10
A female chimpanzee plucks termites from her termiting tool. Some workers think that regional variation in tool use is the product of cultural transmission. (Photograph courtesy of William C. McGrew.)

NUTCRACKING Chimpanzees in the Taï Forest crack open hard-shelled nuts with stone hammers that they pound against other stones and exposed tree roots. This technique is difficult for young chimpanzees to master, and they do not become proficient at cracking nuts for many years. Infants often watch as their mothers crack nuts; mothers often share their hammers and their nut meats with their offspring. One mother watched her five-year-old daughter struggle unsuccessfully to crack open nuts with a hammer stone that she held awkwardly. After several minutes, the mother approached and her daughter gave her the hammer stone. Then the mother very slowly and deliberately rotated the hammer into the correct position. She cracked open several nuts and shared the contents with her daughter. Then the mother left and the daughter picked up the hammer stone, oriented it in the same position as her mother had, and began cracking nuts. This kind of teaching has not been observed in other wild chimpanzee populations and is quite rare even at Taï.

POTATO WASHING This behavior was invented when researchers studying a group of Japanese macaques, whose range included a sandy beach, provisioned them with sweet potatoes. A young female macaque accidentally dropped her sweet potato into the sea as she was trying to rub sand off it. She must have liked the result, as she began to carry all of her potatoes to the sea to wash them (Figure 18.11). Other monkeys followed suit. However, it took other members of the group quite some time to acquire the behavior, and many monkeys never washed their potatoes.

FIGURE 18.11
A young Japanese macaque learned to wash her sweet potatoes, and many members of her group subsequently began to wash their potatoes as well. For many years, this was thought to be a good example of social learning. However, we now think each monkey learned the behavior by trial and error.

Social facilitation and observational learning are similar in that they both can lead to persistent behavioral differences between populations. However, there is an important distinction between the two processes. Social facilitation can only preserve variation in behavior that organisms can learn on their own, albeit in favorable circumstances. Observational learning allows cumulative cultural change. To see the difference, consider the cultural transmission of stone-tool use. Suppose that on her own, in especially favorable circumstances, an early hominid learned to strike rocks together to make useful flakes. Her companions, who spend time near her, would be exposed to the same kinds of conditions, and some of them might learn to make flakes too, entirely on their own. This behavior could be preserved by social facilitation because groups in which tools were used would spend more time in proximity to the appropriate stones. However, that would be as far as it would go. If an especially talented individual found a way to improve the flakes, this innovation would not spread to other members of the group because each individual had to learn the behavior by himself. With observational learning, on the other hand, innovations can persist as long as younger individuals are able to acquire the modified behavior by observing the actions of others. As a result, observational learning can lead to the cumulative evolution of behaviors that no single individual could invent on its own.

Several lines of evidence suggest that social facilitation, not observational learning, is responsible for most cultural traditions in other primates. First, many of the behaviors described in Box 18.5, like potato washing and pod dipping, are relatively simple and could be learned independently by individuals in each generation. Second, new behaviors like potato washing often take a long time to spread through the group, a pace more consistent with the idea that each individual had to learn the behavior on its own. Finally, experiments with capuchin monkeys, known for their ability to use tools and to manipulate objects, suggest that even very clever monkeys like capuchins cannot learn by observation. Primatologist Elisabetta Visalberghi, of the Consiglio Nazionale delle Ricerche in Rome, gave capuchins the opportunity to learn to use a stick to push a peanut out of a horizontal, clear plastic tube. She then allowed a second set of monkeys to watch the skilled monkeys get the highly desired peanut out of the tube. If observational learning were important to this species, we would expect these monkeys to learn the behavior much more rapidly than monkeys who did not have skilled models to imitate. However, observing the skilled monkeys didn't speed up the learning process in the new participants. As Visalberghi and her colleague, Dorothy Fragazy, of Washington State University, put it, "Monkeys don't ape."

We can make a slightly better case for observational learning among apes. It is possible that young chimpanzees, for example, learn to fish for termites and to crack nuts by watching their mothers and imitating their behavior, though this is far from certain. These behaviors seem to be more complicated and harder to learn than the cultural behaviors of monkeys. Psychologist Anne Russon of York University has compiled several anecdotes about the acquisition of human behaviors by orangutans at a rehabilitation center operated by primatologist Biruté Galdikas. Orangutans have extensive contact with humans at the rehabilitation center and observe them closely. At various times, orangutans have stolen rowboats and paddled away down the river or attempted to siphon gas from discarded gas cans. One orangutan followed along behind a worker who was trimming plants growing over the path. The orangutan pulled up plants alongside the path and made his own neat piles in the middle of the path. What is interesting about these anecdotes is that these behaviors are not part of the orangutan's natural repertoire and are unlikely to have been intrinsically rewarding to the orangutans.

Nonetheless, it seems clear that no other primate relies on observational learning to the same extent that humans do, and that their behavior is much less variable from group to group or from region to region than the behavior of humans is.

CULTURE IS AN ADAPTATION

Observational learning is not a by-product of intelligence and social life.

Chimpanzees and capuchins are among the world's cleverest creatures. In nature, they use tools and perform many complex behaviors; in captivity, they can be taught extremely demanding tasks. Chimpanzees and capuchins live in social groups and have ample opportunity to observe the behavior of other individuals, and yet the best evidence suggests that neither chimpanzees nor capuchins make much use of observational learning.

This indicates that observational learning is not simply a by-product of intelligence and having opportunities for observation. Instead, observational learning seems to require special psychological mechanisms. This conclusion suggests, in turn, that the psychological mechanisms that enable humans to learn by observation are adaptations that have been shaped by natural selection because culture is beneficial (Figure 18.12). Of course, this need not be the case. Observational learning could be a by-product of some other adaptation that is unique to humans, such as language. However, given the great importance of culture in human affairs, it is reasonable to think about the possible adaptive advantages of culture.

Psychological mechanisms that allow cumulative cultural change may have been favored by selection because they allowed Pleistocene humans to exploit a much wider range of habitats than any other animal species.

The archaeological record suggests that during the Pleistocene, humans occupied virtually all of Africa, Eurasia, and Australia. Modern hunter-gatherers developed an astounding variety of subsistence practices and social systems. Consider just a few examples. The Copper Eskimos lived in the high Arctic, spending their summers hunting near the mouth of the MacKenzie River and the long dark months of the winter living on the sea ice, hunting seals. Groups were small and heavily dependent on hunting. The !Xõ lived in the central Kalahari, collecting seeds, tubers, and melons, hunting impala and gemsbok, enduring fierce heat, and living without surface water for months at a time. Their small, nomadic bands were linked together in large band clusters organized along male kinship lines. The Chumash lived on the southern California coast, gathering shellfish and seeds and fishing the Pacific from great plank boats. They lived in large permanent villages with pronounced division of labor and extensive social stratification.

The fact that the !Xõ could acquire the knowledge, tools, and skills necessary to survive the rigors of the Kalahari is not so surprising—many other species can do the same. What is amazing is that the same brain that allowed the !Xõ to survive in the Kalahari also permitted the Copper Eskimo to acquire the very different knowledge, tools, and skills necessary to live on the tundra and ice north of the Arctic Circle, and the Chumash to acquire the skills necessary to cope with life in crowded, hierarchical Chumash settlements. No other animal occupies a comparable range of habitats or utilizes a comparable range of subsistence techniques and social structures. For example, savanna baboons, the most widespread primate species, are limited to Africa and Arabia, and the diet, group size, and social systems of these far-flung baboon populations vary to a much smaller degree than the diet, group size, and social systems of human hunter-gatherers.

Humans can adapt to a wider range of environments than baboons because they acquire knowledge about the environment culturally, not genetically. Baboons cannot acquire behavior by observational learning because their behavior is limited to things that individuals can learn on their own. As we have seen, general-purpose learning mechanisms are costly and inefficient compared with mechanisms that work in a relatively narrow range of environments or solve limited kinds of problems. Thus, it is plausible that the innate psychological machinery of baboons allows them to adapt only to a relatively narrow range of environments. In contrast, because humans *can* acquire behavior by observational learning, cultural variation can accumulate gradually, and humans are often able to make use of skills and knowledge that cannot be acquired by any single individual. In this way, less specialized psychological mechanisms allow humans to adapt to a wider range of environments than baboons can.

Culture can lead to evolutionary outcomes not predicted by ordinary evolutionary theory.

FIGURE 18.12

Infants are prone to spontaneously imitating behaviors they observe. Here, a 13-month-old infant flosses her two teeth.

FIGURE 18.13

Cultural evolution may permit the spread of ideas and behaviors that do not contribute to reproductive success. Dangerous sports, like rock climbing, may be examples of such behaviors.

Once culture becomes established, it is like a second system of inheritance. We acquire genes from our parents, and we acquire culturally transmitted beliefs and values from parents, relatives, friends, and, perhaps, even from professors. To emphasize this similarity, biologist Richard Dawkins refers to culturally transmitted beliefs and values as **memes**.

The logic of natural selection applies to memes in the same way that it does to genes. Memes compete for our memory and for our attention, and not all memes can survive. Some memes are more likely to survive and be transmitted than others; memes are heritable, often passing from one individual to another without major change. As a result, some memes spread and others are lost.

However, the fact that memes are subject to natural selection does not mean that genetically adaptive memes will spread. Because the rules of cultural transmission are different from the rules of genetic transmission, the outcome of selection on memes can be different from the outcome of selection on genes. The rules of genetic transmission are simple. With some exceptions, every gene that an individual carries in her body is equally likely to be incorporated into her gametes, and the only way those genes will be transmitted is if she produces children who carry those gametes. Thus, only genes that increase reproductive success will spread. Cultural transmission, by contrast, is Byzantine in its complexity. Memes are acquired and transmitted piecemeal throughout an individual's entire life, not just from parents to offspring, but from grandparents, siblings, friends, coworkers, teachers, and even completely impersonal sources like books, television, and the Internet. Unlike genes, which are transmitted with little modification from parents to offspring, memes may spread along many different pathways. Ideas about dangerous hobbies like rock climbing or heroin use can spread from friend to friend, even though these behaviors are likely to reduce the chance of reproducing successfully (Figure 18.13). Beliefs about heaven and hell can spread from priest to parishioner, even if the priest is celibate. To add to the complexity, much heritable cultural variation accumulates among groups of people who form clans, fraternities, business firms, religious sects, or political parties. Memes that improve, say, quality assurance, can spread through a business community because they cause firms to earn profits, grow, and spawn new firms with the same methods of quality assurance.

The fact that culture can lead to outcomes not predicted by conventional evolutionary theory does not mean that human behavior has somehow transcended biology. The idea that culture is separate from biology is a popular misconception that cannot withstand scrutiny. Culture cannot transcend biology because it is as much a part of human biology as bipedal locomotion. Culture is generated from organic structures in the brain that were produced by the processes of organic evolution. However, cultural transmission leads to novel evolutionary processes. Thus, to understand the whole of human behavior, evolutionary theory must be modified to account for the complexities introduced by these, as yet poorly understood, processes.

The fact that culture can lead to outcomes that would not be predicted by conventional evolutionary theory does not mean that ordinary evolutionary reasoning is useless. The fact that there are processes that lead to the spread of risky behaviors like rock climbing does not mean that these are the only processes that influence cultural behavior. Earlier in this chapter we saw that it is likely that many aspects of human psychology have been shaped by natural selection so that people learn to behave adaptively. We love our children, prefer not to mate with relatives, and think adeptly about social contracts. There is every reason to suspect that these predispositions play an important role in determining which memes spread and which don't. To the extent that this observation is true, ordinary evolutionary reasoning will be useful for understanding human behavior.

Human Behavioral Ecology

Human behavioral ecologists view humans as rational actors who seek to maximize their fitness.

The workings of the human mind are not a central problem for most social scientists. Instead, they want to understand human behavior and social organization. They ask, why do some people subsist via hunting and gathering while others rely on agriculture? Why are some societies egalitarian while other societies are hierarchical? Why are some societies polygynous and others monogamous? Although in principle we might answer such questions by identifying special-purpose modules in the brain that evolved to solve specific adaptive problems faced by Pleistocene foragers, this is not a practical alternative at present. Evolutionary psychology is a young discipline, and we know very little about how natural selection has shaped specific aspects of human psychology.

Some researchers have, instead, adopted the same kinds of tactics that evolutionary biologists use when they want to understand particular aspects of the morphology and behavior of other organisms. As we have seen throughout this book, the assumption that phenotypes have been shaped by natural selection provides a powerful tool for understanding organisms. It can help to explain why different primates have very different kinds of teeth and why male monkeys care for their offspring in some pair-bonded groups, but not in other types of groups. However, we know from Chapter 3 that there are several reasons why evolution may not produce adaptations in every case: selection acting on one character may affect other characters, genetic drift may lead to maladaptation, or constraints may prevent adaptation from occurring. To be sure that adaptive reasoning is justified in any particular case, a biologist would need to know enough about the genetics, development, and history of the organism of interest to discount maladaptive processes. However, biologists almost never know much about these things. Many biologists simply ignore these complications and assume that observed phenotypes are adaptive, a research tactic that Oxford biologist Alan Grafen calls the **phenotypic gambit.** Biologists use the phenotypic gambit, knowing that it may not be correct in every case, because experience in many areas of biology suggests that adaptive reasoning is often very useful.

A group of researchers called human behavioral ecologists have applied the phenotypic gambit to the human species. Their theory assumes that contemporary humans act as if they are attempting to maximize their genetic fitness. The assumption that humans are rational actors who maximize their own selfish interests is widely used in the social sciences. Economists have used this idea to build an elaborate and successful theory that explains how people behave in market settings. Economists and other social scientists have extended rational-actor models to account for political behavior, marriage, decisions about family size, and many other kinds of behavior in nonmarket settings. While these applications are controversial in sociology and political science, there is little doubt that the rational-actor approach is important in all of the social sciences. Human behavioral ecology is simply rational-actor theory that goes one step further. It predicts that peoples' interests are defined in evolutionary terms—humans want to maximize their fitness. As we will see, this step provides human behavioral ecologists with a rich source of hypotheses about human behavior, particularly for traditional societies in which markets play a relatively minor role.

Because animals must acquire enough food to meet their energy needs, natural selection tends to favor adaptations that increase foraging efficiency.

No matter what kind of food an animal eats, or how it obtains its food, natural selection usually favors foraging efficiency. Sometimes, morphological adaptations enhance foraging abilities. For example, some folivorous monkeys have complex stomachs that increase their ability to extract nutrients from leaves. Foraging efficiency, measured in terms of the amount of energy obtained per unit of time, may also be enhanced by reducing the amount of time it takes to locate food, by selecting foods that have high nutritional value and low processing time, or by abandoning food patches when they are depleted.

In behavioral ecology, the evolution of these types of adaptations is the domain of **optimal foraging theory.** This theory is based on the assumption that natural selection shapes foraging behavior so as to maximize the amount of energy gained per unit of time. Of course, this assumption is not exactly true because getting food is not the only problem facing animals. For example, primates might be more efficient if they foraged alone, but they forage in groups to reduce the risk of predation. However, foraging efficiency is so important that it is often reasonable to ignore other factors. Optimal foraging theory generates quantitative predictions about foraging behavior that can be tested in real-life situations. This body of theory has been applied to the Aché, a hunter-gatherer people living in the forests of Paraguay.

Optimal foraging theory has been used to understand the economy of the Aché.

Until the mid-1970s, the Aché were nomadic foragers who subsisted entirely on resources that they obtained in the forests of Paraguay. Like many other indigenous peoples, the Aché's lives have been radically altered by contact with Western culture. They have been encouraged to settle near missions, to grow crops, and to raise livestock. Nonetheless, small bands of Aché continue to make periodic foraging trips into the forest, much as they did before European contact. The Aché are proficient hunters (Figure 18.14) who take many animal species, including armadillos, monkeys, and peccaries (small piglike mammals). Men sometimes head off into the forest together, fanning out in different directions to search for game, but remain close enough to call out when they spot prey and to coordinate the chase and capture. Men also collaborate in locating and extracting honey, which is greatly prized. Women make camp, tend children, collect wood for fires, help in hunts, and gather plant foods (Figure 18.15). In the forest, the Aché obtain about 3600 calories per person each day, and about 70% of the calories come from meat.

As we described in Chapter 12, a group of anthropologists from the University of Utah (that included Kristin Hawkes, Kim Hill, Magdalena Hurtado, and Hillard Kaplan) has made careful studies of the Aché's subsistence ecology. They monitored the time that hunters spent in search and pursuit of prey and the amount of time that they needed to process

FIGURE 18.14
Aché men are skilled hunters, but they do not always forage optimally. If they did, they would spend less time hunting and more time processing palm starch. (Photograph courtesy of Kim Hill.)

HUMAN BEHAVIORAL ECOLOGY

their kills. They also kept track of the amount of time gatherers spent searching for and processing plant foods. They tried to weigh all of the food that each member of the foraging group obtained. Using this information, along with information about the caloric values of various foods, they were able to estimate the number of calories obtained per hour of effort, as well as the rate of return for particular foods and for individual foragers.

Optimal foraging theory predicts that foragers will exploit only those resources that provide a rate of return that is greater than the overall return rate for all resources.

On an average day, a forager can expect to obtain a given number of calories per hour. This is based on all the food she obtains and all the time that it takes to find, pursue, and process these food items. Now, as the forager moves through the forest, she encounters a potential food item, say, a palm tree. Should she continue walking, or should she begin to process palm to extract the starch? If more productive resources are nearby, then she should keep walking. But how does she know whether there are more productive resources in the vicinity? Her best estimate of the return for alternative foods is simply the average rate of return for foraging. Thus, her decision will be based on the difference between the rate of return from the resource she has just found and the overall mean rate of return. If the difference between these rates is positive, then she will on average do best by beginning to process palm starch. If the difference is negative, then she will be better off to continue her search for food.

With two important exceptions, Aché behavior conforms to the predictions of optimal foraging theory.

The Aché concentrated their foraging efforts on resources that generated a higher rate of return than the mean (Figure 18.16). They tended to take the most productive resources most often. However, the University of Utah researchers discovered two major discrepancies between the predictions of optimal foraging theory and the Aché's foraging behavior. First, the men routinely ignore plant foods, even the ones that generate relatively high rates of return (Figure 18.16a). Men could greatly increase the number of calories they obtain per hour if they devoted themselves to extracting and processing fiber from the trunks of palm trees. Instead, men hunt.

Second, women routinely ignore resources, like honey and meat, that bring high rates of return, and concentrate their efforts on plant foods with lower return rates (Figure 18.16b). These observations clearly violate the predictions of the optimal foraging theory model.

Evolutionary theory suggests reasons why Aché men hunt rather than gather and why women gather rather than hunt.

The optional foraging analysis clearly shows that if men were trying to maximize the rate of acquiring calories, they would give up their bows and shotguns and become full-time gatherers. And the women would occasionally pass up palms to hunt game. Evolutionary theory provides some clues to understand why men and women don't forage optimally.

One possibility is that the optimal foraging models ignore things that matter to Aché foragers. For example, the model assumes all calories are of equal value. How-

FIGURE 18.15
Aché women spend much of their time gathering plant foods and processing palm starch. (Photograph courtesy of Kim Hill.)

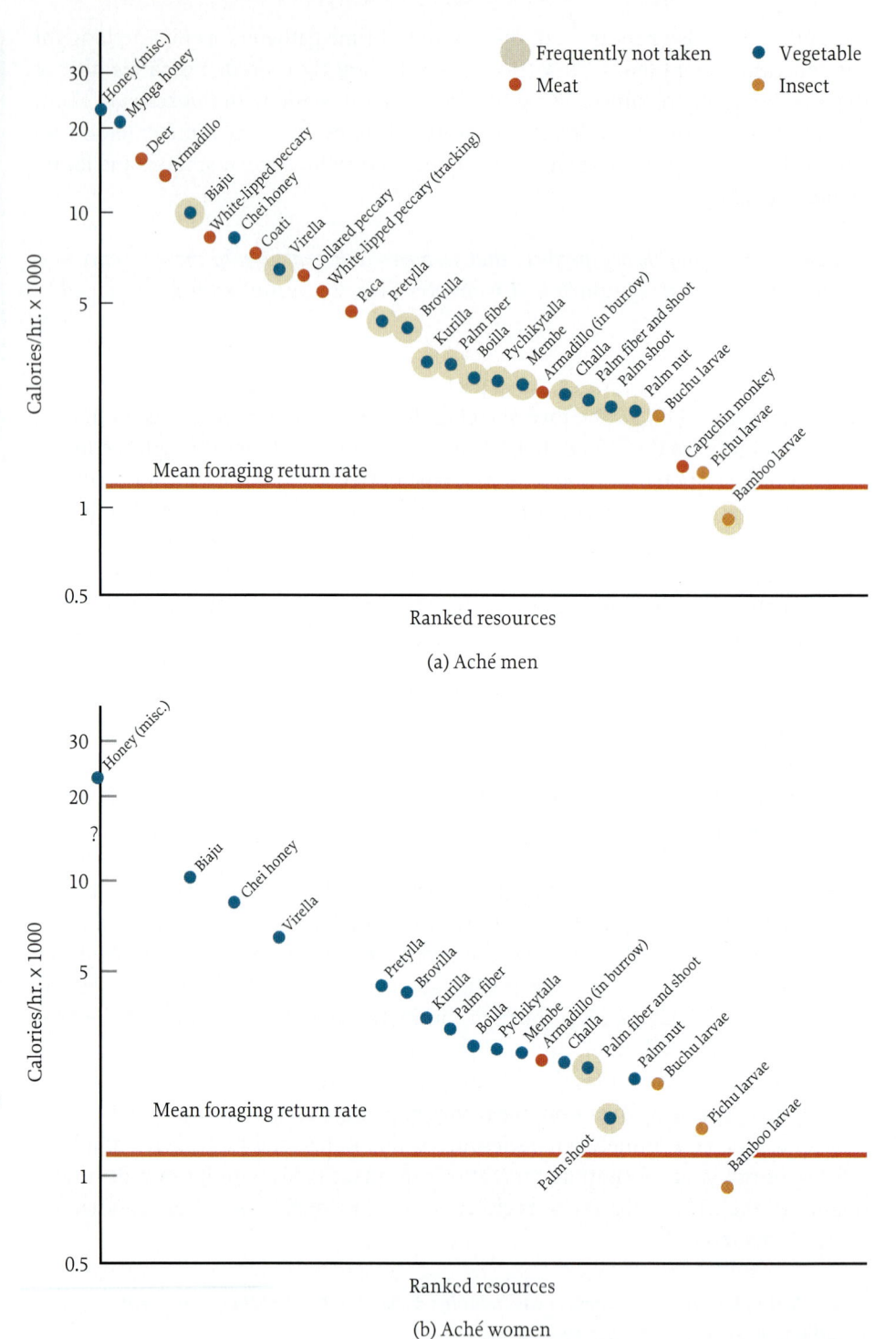

FIGURE 18.16

The Aché gather a variety of plant foods and hunt wild game. Their foraging behavior generally conforms to optimal foraging models, as they concentrate on resources that bring higher rates of return (calories/hour) than the mean. (a) However, men routinely ignore plant foods that would increase their foraging efficiency; and (b) women ignore honey and meat, which would increase their foraging efficiency.

ever, calories obtained from meat may be more valuable than calories obtained from plant foods. Meat provides both protein and fat and may be of greater nutritional value than equivalent amounts of carbohydrates. The problem with this explanation is that the Aché eat much more meat than most people do. They obtain nearly 70% of their calories from meat, and their consumption of protein and fat greatly exceeds the body's daily requirements for these nutrients. Thus, Aché men could hunt less and gather more, without jeopardizing their nutritional status.

Hunting may also increase men's reproductive success because better hunters get social payoffs that increase their reproductive success. There are substantial differences among men in foraging efficiency. In general, young and old men forage less efficiently than men in their prime do. Among middle-aged men, some are better hunters than others, bringing back substantially more meat per unit of time spent hunting. These differences are apparently the result of skill, not luck, as the differences among men persist over several years. Aché men's reproductive success is correlated with their foraging efficiency; the men with the highest return rate had the largest number of surviving children by their wives and the greatest number of illegitimate children. Interestingly, this correlation does not arise because these men are able to provide more food for their own children than other men do. The Aché share their kills with all members of their foraging party, and the hunter's family does not get a disproportionate share of his kills. Instead, the University of Utah group speculated that successful hunters may gain reproductive advantages in other ways. As an inducement for successful hunters to remain in the group, other members of the band may treat a good hunter's children better, being more accommodating when they are ill, grooming them, or protecting them from the many hazards in the forest. Members of the group benefit from the presence of a successful hunter who shares his kills; the hunter benefits by increasing the chance that his children will survive.

This example illustrates why evolutionary theory is a rich source of hypotheses about human behavior. The research team started with a simple picture of the Aché economy that generated several detailed predictions about foraging patterns. They collected the requisite data, and it turned out that much of Aché foraging behavior conformed to the model, suggesting that the Aché forage efficiently. However, the model failed to account for why men don't process palms and why women don't hunt. Because evolutionary theory specifies what people should value, it allows us to determine what was left out of the model, such as other nutrients. By modifying the model to include these factors, we can generate new predictions and make new observations to test these predictions. This cycle of theory and observation is a potent process for generating a richly textured picture of the Aché subsistence economy.

Further Reading

Barkow, J., L. Cosmides, and J. Tooby, eds. 1992. *The Adapted Mind, Evolutionary Psychology and the Generation of Culture.* Oxford University Press, New York.
Berlin, B., and P. Kay. 1991 (1969). *Basic Color Terms.* University of California Press, Berkeley. (The 1991 paperback edition has a useful supplementary review of developments since the original publication in 1969.)
Boyd, R., and P. Richerson. 1985. *Culture and the Evolutionary Process.* University of Chicago Press, Chicago.
Brown, D. 1991. *Human Universals.* McGraw-Hill, New York.
Smith, E. A., and B. Winterhalder. 1992. *Evolutionary Ecology and Human Behavior.* Aldine de Gruyter, Hawthorne, N.Y.

Study Questions

1. Much of the behavior of all primates is learned. Nonetheless, we have suggested many times that primate behavior has been shaped by natural selection. How can natural selection shape behaviors that are learned?

2. Many of the things that we do are consistent with general predictions derived from evolutionary theory. We love our children, help our relatives, and avoid sex with close kin. But there are also many aspects of the behavior of members of our own society that seem unlikely to increase individual fitness. What are some of these behaviors?
3. In some species of primates there seems to be an aversion to mating with close kin. The aversion seems to be stronger for females than for males. Why do you think this might be the case? Under what conditions would you expect this gender difference to disappear?
4. The verb *to ape* means "to copy or imitate." Is its meaning consistent with what we now know about the learning processes of other primates?
5. Evolutionary psychologists and human behavioral ecologists take a very different view of the adaptive value of contemporary human behavior. Explain how researchers working in each of these areas view contemporary behavior. What aspects of modern behavior seem to be consistent with each point of view?

CHAPTER 19

Human Mate Choice and Parenting

- THE PSYCHOLOGY OF HUMAN MATE PREFERENCES
- SOME SOCIAL CONSEQUENCES OF MATE PREFERENCES
  - KIPSIGIS BRIDEWEALTH
  - NYINBA POLYANDRY
- RAISING CHILDREN
  - CHILD ABUSE AND INFANTICIDE
  - ADOPTION
  - FAMILY SIZE
- IS HUMAN EVOLUTION OVER?

You should be convinced by now that every aspect of the human phenotype is the product of evolutionary processes. We humans are large, nearly hairless, bipedal primates, having grasping hands and large brains because natural selection, mutation, and genetic drift made us that way over the last 10 million years. The same evolutionary processes have molded the psychological mechanisms that influence human behavior, causing us to behave differently than other primates in some contexts. There is simply no other explanation.

However, the fact that modern humans are the product of evolutionary processes does not necessarily mean that evolutionary theory will help us understand contemporary human behavior. Constraints on adaptation, which we discussed in Chapter 3, may have predominated in the evolution of human behavior, limiting the usefulness of adaptive reasoning. However, in Chapter 18 we saw that evolutionary theory has provided useful insights into certain aspects of human psychology, social organization, and culture. In this chapter, we present several more examples that show how evolutionary thinking can help us understand the day-to-day behavior of humans in modern societies.

In choosing these examples, we have focused on reproductive behavior because mating and parenting strongly affect fitness. The unfortunate person who chooses a mate who is lazy, infertile, or unfaithful is likely to have many fewer children than the person whose spouse is healthy, hardworking, and faithful. The inattentive or abusive parent will have fewer children than the parent who carefully nurtures his or her offspring. Because reproductive decisions are likely to have a marked effect on fitness, there is good reason to expect that our psychology has been shaped by natural selection to improve the chances of making good choices. We begin by presenting research on the psychology of mate preferences, and then consider two examples that illustrate how such preferences work themselves out in real social settings. We then turn to studies that illustrate how evolutionary theory can help us to understand the range of variation in human parenting strategies. We do not have enough space to provide a comprehensive analysis of these topics. Instead, we focus on certain examples in which evolutionary theory provides novel and fundamental insights into human behavior.

The Psychology of Human Mate Preferences

Marry

Children—(if it Please God)—Constant companion, (& friend in old age) who will feel interested in one,—object *to be* beloved and played with. better than a dog anyhow.—Home, & someone to take care of house—Charms of music & female chit-chat.—These things good for one's health.—*but terrible loss of time.*—

My God, it is intolerable to think of spending one's whole life, like a neuter bee, working, working, & nothing after all.—No, no won't do.— Imagine living all one's day solitary in smoky dirty London house.— Only picture to yourself nice soft wife on a sofa with good fire, & books, & music perhaps—Compare this vision with the dingy reality of Grt. Marlbro St.

Marry—Mary—Marry Q.E.D.

Not Marry

Freedom to go where one liked—choice of Society & *little of it.*—Conversation of clever men at clubs—Not forced to visit relatives, & to bend in every trifle.—to have the expense & anxiety of children— perhaps quarelling—**Loss of time.**—cannot read in the Evenings—fatness & idleness—Anxiety & responsibility—less money for books &c—if many children forced to gain one's bread.—(But then it is very bad for ones health to work too much)

Perhaps my wife wont like London; then the sentence is banishment & degradation into indolent, idle fool.

[From p. 444 in F. Burkhardt and S. Smith, eds., 1986, *The Correspondence of Charles Darwin*, Vol. 2, 1837–1843, Cambridge University Press, Cambridge, U.K.]

FIGURE 19.1
Charles Darwin courted and married his cousin, Emma Wedgwood. This portrait was painted when Emma was 32, just after the birth of her first child.

These are the thoughts of 29-year-old Charles Darwin, recently returned from his five-year voyage on the *Beagle*. Eventually, Darwin married his cousin, Emma, the daughter of Josiah Wedgwood, the progressive and immensely wealthy manufacturer of Wedgwood china (Figure 19.1). They were, by all accounts, a devoted couple. Emma bore Charles 10 children and nursed him through countless bouts of illness. Charles toiled over his work and astutely managed his investments, parlaying his modest inheritance and his wife's more substantial one into a considerable fortune.

Darwin's frank reflections on marriage were very much those of a conventional, upper-class, Victorian gentleman. But people of every culture, class, and gender have faced the problem of choosing mates. Sometimes people choose their own mates, and at other times parents arrange their children's marriages. But everywhere, people care about the kind of person they will marry.

Evolutionary theory generates a number of testable predictions about the psychology of human mate preferences.

David Buss, a psychologist at the University of Texas, conducted a series of studies designed to assess whether predictions derived from evolutionary theory fit patterns of human mate preferences in contemporary human societies. Buss reasoned that the psychology underlying contemporary human mate preferences was shaped during the Pleistocene when humans were hunters and gatherers. Using this approach, he generated the following predictions about evolved mate preferences:

- *Women prefer males who are able and willing to provide resources.* In the environment of evolutionary adaptedness, it may have been difficult for anyone to accumulate resources, but some men were probably better providers than others. Women may have preferred men who could provide the most resources for themselves and for their offspring. As foraging gave way to horticulture, and then to agriculture, the capacity for accumulating resources increased dramatically. In many contemporary societies, there are great disparities in wealth between individuals. Females' preference for good providers may now be expressed as a preference for wealthy men.
- *Males prefer females who will reproduce successfully.* A man's reproductive success will depend partly on the reproductive success of his partner. This, in turn, will depend on his partner's reproductive value and fertility. A woman's **reproductive value** (her expected contribution to future generations) peaks before she reaches 20, steadily declines until she reaches menopause, and then drops to zero. Female **fertility,** or birth rate, is highest in the early 20s. In general, men are expected to prefer younger women to older women. Some researchers have suggested that cross-cultural standards of beauty reflect an evolved preference for physical traits that are generally associated with youth, such as smooth skin, good muscle tone, and shiny hair. Therefore, men are expected to value beauty in prospective mates.
- *Both men and women value fidelity, but men value it more highly than women.* When both parents provide care for offspring, both males and females have reason to value fidelity. Men should not want to invest resources in another man's children, and women should not want their mates to divert resources to another female's offspring. However, the value of fidelity is likely to be greater for men than for women because men can be cuckolded. A man can only be sure that he is the father of all his wife's children if he is certain she has been completely faithful during their marriage, while a woman knows that she is the mother of all her children, no matter what her husband does. Thus, men are likely to value fidelity more than women do.

Although Buss focused mainly on traits that men and women are likely to value differently, there are also traits that we might expect both sexes to consider important. Since parental investment lasts for many years, we would expect both men and women to value traits in their partners that help them sustain their relationships.

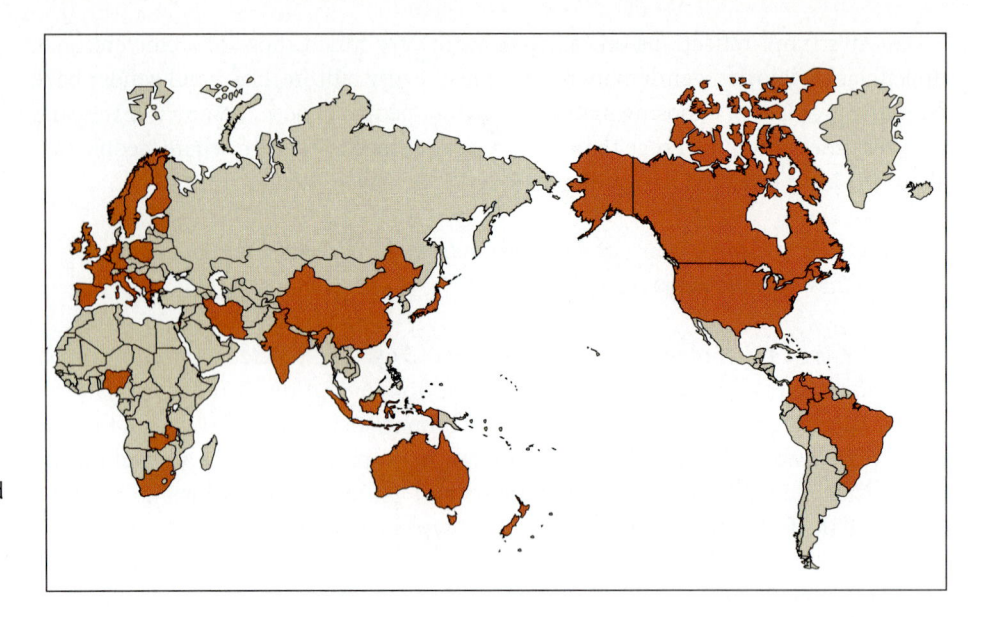

FIGURE 19.2

People in 33 countries (red) were surveyed about the qualities of an ideal mate.

Both are likely to value personal qualities such as compatibility, kindness, helpfulness, and tolerance.

Buss tested the importance of these factors to men and women across the world. He enlisted colleagues in 33 countries to administer standardized questionnaires about the qualities of desirable mates to more than 10,000 men and women. Most of the data were collected in Western industrialized nations, and most samples represent urban populations within those countries (Figure 19.2). In the questionnaires, people were asked to rate 18 traits of potential mates, such as good looks, financial prospects, compatibility, and so on, according to their desirability. Respondents were also asked about their preferred age at marriage, the preferred age difference between themselves and their spouse, and the number of children that they wanted to have. The results provide many interesting insights about human mate choice, but we will limit our account to a few key points.

People generally care most about the personal qualities of their mates.

People around the world rate mutual attraction or love above all other traits (Table 19.1). The next most highly desired traits are personal attributes, such as dependability, emotional stability and maturity, and a pleasing disposition. Good health is the fifth most highly rated trait for men and the seventh for women. Good financial prospect is the 13th most highly rated trait by men and the 12th by women. Good looks are rated 10th by men and 13th by women. It is interesting, and somewhat surprising, that neither sex seems to value chastity highly. Perhaps this is because people were asked to evaluate the desirability of sexual experience before marriage (that is, virginity), not fidelity during their marriage.

Men and women show the differences in mate preferences predicted by evolutionary theory.

You can see that there is a high degree of concordance in the rankings of the scores that men and women assign to the traits in Table 19.1. For example, men and women both rate love more highly than any other trait, and give the lowest ratings to political

TABLE 19.1 People from more than 30 countries around the world were asked to rate the desirability of a variety of traits in prospective mates. The rankings of the values assigned to each trait, on average, by men and women in countries around the world are given here. Subjects were asked to rate each trait from 0 (irrelevant or unimportant) to 3 (indispensable). Thus, high ranks (low numbers) represent traits that were generally thought to be important. (From Table 4 in D. M. Buss et al., 1990, International preferences in selecting mates: a study of 37 cultures, *Journal of Cross-Cultural Psychology* 21:5–47.)

Trait	Ranking by Males	Ranking by Females
Mutual attraction-love	1	1
Dependable character	2	2
Emotional stability and maturity	3	3
Pleasing disposition	4	4
Good health	5	7
Education and intelligence	6	5
Sociability	7	6
Desire for home and children	8	8
Refinement, neatness	9	10
Good looks	10	13
Ambition and industrious	11	9
Good cook and housekeeper	12	15
Good financial prospect	13	12
Similar education	14	11
Favorable social status or rating	15	14
Chastity*	16	18
Similar religious background	17	16
Similar political background	18	17

*Chastity was defined in this study as having no sexual experience before marriage.

and religious affinities. Even though the ranking of the scores assigned to these traits is similar for men and women, Buss emphasized that there are still important and consistent differences between men and women in how desirable each gender thinks these traits are. Buss found that a person's gender had the greatest effect on their ratings of the following traits: "good financial prospect," "good looks," "good cook and housekeeper," "ambition and industrious." As the evolutionary model predicts, women value good financial prospects and ambition more than men do, while men value good looks more than women do. Gender has a smaller and somewhat less uniform effect on ratings of chastity. In 23 populations, men value chastity significantly more than women do, while in the remaining populations men and women value chastity equally. There are no populations in which women value chastity significantly more than men do. Men and women also differ about the preferred age differences between themselves and their spouses. Men invariably want to marry women who are younger than themselves, while women want to marry men who are older than they are.

> *Buss's analysis shows that culture exerts a stronger influence on people's mate preferences than gender does.*

The country of residence has a greater effect than gender does on variation in all of the 18 traits in Table 19.1, except for "good financial prospects." This means that knowing where a person lives tells you more about what he or she values in a mate than knowing the person's gender. Of the 18 traits, chastity (defined in Buss's study as no sexual experience before marriage) shows the greatest variability among populations. In Sweden, men rate chastity at 0.25, and women rate it at 0.28 on a scale of 0 (irrelevant) to 3 (indispensable). In contrast, Chinese men rate chastity 2.54, and Chinese women rate it 2.61 (Figure 19.3). Buss and his colleagues interpreted these results to mean that there is more similarity between men and women from the same population than there is among members of each sex from different populations.

This result illustrates an important point: evolutionary explanations that invoke an evolved psychology and cultural explanations that are based on the social and cultural milieu are not mutually exclusive. Buss's findings suggest there are cross-cultural uniformities in people's mate preferences that are the result of evolved psychological mechanisms. People want to marry kind, caring, trustworthy people. Men want to marry younger women and women want to marry older men. But this is not the whole story. Buss's data also suggest that human mate preferences are strongly influenced by the cultural and economic environment in which we live. Ultimately, culture also arises out of our evolved psychology, and the cultural variation in mate preferences that Buss observed must therefore also be explicable in evolutionary terms. However, the way our evolved psychology shapes the cultures in which we live is complicated and poorly understood, and many interesting questions remain unresolved. Evolution-

(a) Chastity

(b) Good financial prospect

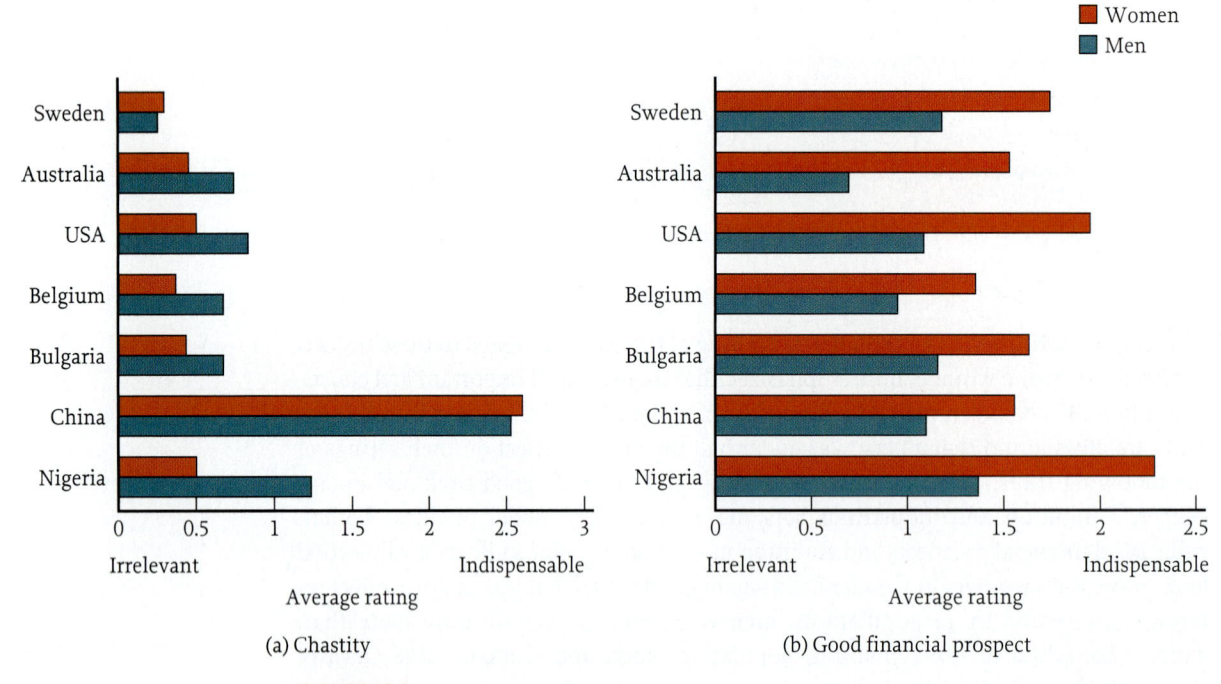

FIGURE 19.3

Culture accounts for substantial variation in mate preferences. The average ratings given by men and women in several countries surveyed are shown for (a) the trait with the highest interpopulation variability ("chastity") and (b) the trait with the lowest interpopulation variability ("good financial prospect").

ary theory does not yet explain, for example, why chastity is essential in China but undesirable in Sweden.

Some Social Consequences of Mate Preferences

You might wonder how people's preferences for certain kinds of traits in prospective mates influence their actual decisions and choices about marriage partners. In this section, we describe the findings from one ethnographic study that suggest these kinds of preferences actually influence people's behavior in social situations, and consequently shape the societies in which they live.

KIPSIGIS BRIDEWEALTH

Evolutionary theory explains marriage patterns among the Kipsigis, a group of East African pastoralists.

Among the Kipsigis, a group of Kalenjin-speaking people who live in the Rift Valley Province of Kenya, women usually marry in their late teens, men usually marry for the first time when they are in their early 20s, and it is common for men to have several wives, a practice called **polygyny.** As in many societies, the groom's father makes a **bridewealth** payment to the father of the bride at the time of marriage. The payment, tendered in livestock and cash, compensates the bride's family for the loss of her labor and gives the groom rights to her labor and the children she bears during her marriage. The amount of the payment is settled through protracted negotiations between the father of the groom and the father of the bride. The average bridewealth consists of six cows, six goats or sheep, and 800 Kenyan shillings. This is about one-third of the average man's cattle holdings, one-half of his goat and sheep herd, and two months' wages for men who hold salaried positions. Since men marry polygynously, there is a competition over eligible women. Often, the bride's father entertains several competing marriage offers before he chooses a groom for his daughter. The prospective bride and groom have little voice in the decisions their fathers make.

Anthropologist Monique Borgerhoff Mulder at the University of California, Davis, reasoned that Kipsigis bridewealth payments provide a concrete index of the qualities that each party values in prospective spouses. The groom's father is likely to prefer a bride who will bear his son many healthy children. His bridewealth offer is expected to reflect the potential reproductive value of the prospective bride. The groom's father is also expected to prefer that his son marry a woman who will devote her labor to his household. Kipsigis women who remain near their own family's households are likely to be called on to help their mothers with the harvest and to assist their mothers in childbirth. Thus, the groom's father may prefer a woman whose natal family is distant from his son's household. The bride's father is likely to have a different perspective on the negotiations. Since wealthy men can provide their wives with larger plots of land to farm and more resources, the bride's father is expected to prefer that his daughter marry a relatively wealthy man. At the same time, since the bride's family will be deprived of her labor and assistance if she moves far away from their land, the bride's father is likely to prefer a groom who lives nearby. The fathers of the bride and groom are expected to weigh the costs and benefits of prospective unions in their negotiations over bridewealth payments. For example, while the bride's father may prefer a high bridewealth payment, he may settle for a lower payment if the

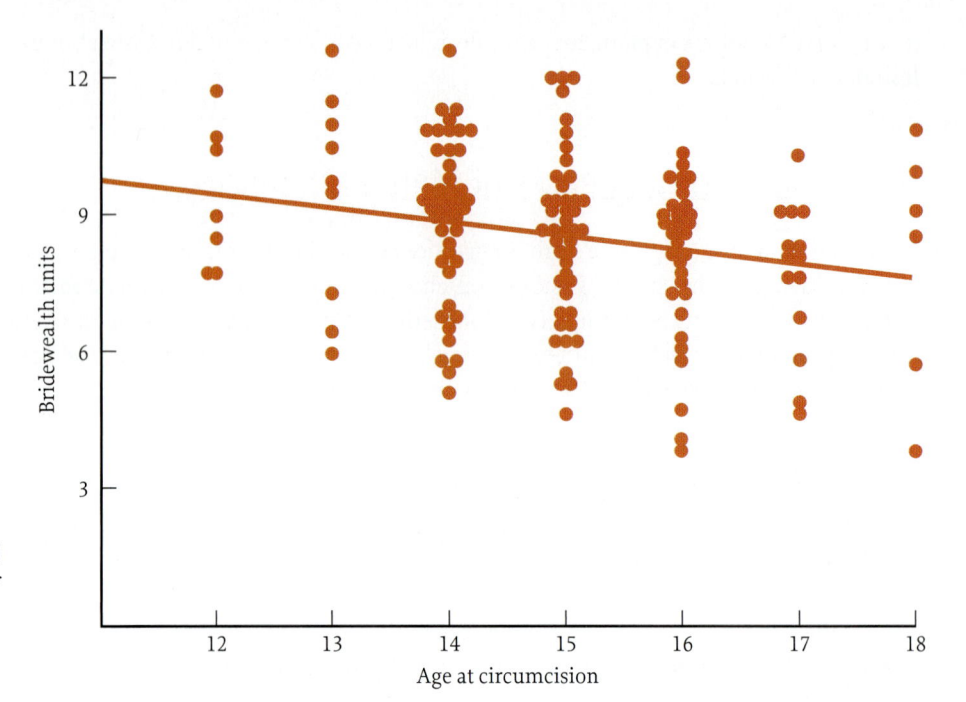

FIGURE 19.4

Kipsigis girls who mature early fetch larger bridewealth payments than older girls do. Among the Kipsigis, girls undergo circumcision (removal of the clitoris) within a year of menarche. The largest bridewealths were paid for the girls who underwent menarche and circumcision at the youngest ages. Bridewealth is transformed into standardized units to account for the fact that the value of livestock varies over time.

groom is particularly desirable. In order to determine whether these preferences affected bridewealth payments, Borgerhoff Mulder recorded the number of cows, sheep, goats, and the amount of money that each groom's family paid to the bride's family.

Plump women whose menarche occurred at an early age fetched the highest bridewealth payments.

Borgerhoff Mulder found that bridewealth increased as the bride's age at the time of **menarche** (her first menstruation) decreased (Figure 19.4). That is, the highest bridewealths were paid for the women who were youngest when they first menstruated (Figure 19.5). Among the Kipsigis, age at menarche is a reliable index of women's reproductive potential. Kipsigis women who reach menarche early have longer reproductive life spans, higher annual fertility, and higher survivorship among their offspring than do women who mature at later ages.

Borgerhoff Mulder wondered how a man assessed his prospective bride's reproductive potential, since men often do not know their bride's exact age, nor her age at reaching menarche. One way to be sure of a woman's ability to produce children would be to select one who had already demonstrated her fertility by becoming pregnant or producing a child. However, bridewealths for such women were typically lower than bridewealths for women who had never conceived. Instead, bridewealth payments were associated with the physical attributes of women. The fathers of brides who were considered by the Kipsigis to be plump received significantly higher bridewealth payments than did the fathers of brides considered to be skinny. The plumpness of prospective brides may be a reliable correlate of the age at menarche, since

FIGURE 19.5

These Kipsigis women are eligible for marriage. Their fathers will negotiate with the fathers of their prospective husbands over bridewealth. (Photograph courtesy of Monique Borgerhoff Mulder.)

menarcheal age is determined partly by body weight. Plumpness may also be valued because a woman's ability to conceive is partly determined by her nutritional status.

Bridewealth payments are also related to the distance between the bride's home and the groom's home; the farther she moves, the less likely she is to provide help to her mother, and the higher the bridewealth payment. However, there is no relationship between the wealth of the father of the groom and the bridewealth payment. The bride's father does not lower the bridewealth payment to secure a wealthy husband for his daughter. Although this finding was unexpected, Borgerhoff Mulder suggested that it may be related to the fact that differences in wealth among the Kipsigis are unstable over time. A wealthy man who has large livestock herds may become relatively poor if his herds are raided, decimated by disease, or diverted to pay for another wife. While land is not subject to these vicissitudes, the Kipsigis traditionally have not held legal title to their lands.

Nyinba Polyandry

Polyandry is rare in humans and other mammals.

Since mammalian females are generally the limiting resource for males, males usually compete for access to females. As a result, polygyny is much more common among mammalian species than polyandry. As you may recall from Chapter 6, polyandry is a mating system in which one female is paired with two or more males. Polyandry is rare because, all other things being equal, males who share access to a single female produce fewer offspring than males who maintain exclusive access to one or more females. The reproductive costs of polyandry may be somewhat reduced if brothers share access to a female, an arrangement called **fraternal polyandry.** Thus, in accordance with Hamilton's rule, it is plausible to assume that natural selection has shaped human psychology so that a man is more willing to care for his brother's offspring than for children who are unrelated to him.

Polyandrous marriage systems are as rare among human societies as they are in other mammalian species. In a sample of 862 societies compiled by anthropologist George Peter Murdock, polygyny is the ideal form of marriage in 83% of societies, 16% are exclusively monogamous, and only 0.5% are polyandrous.

Polyandrous marriage occurs among several societies in the Himalayas.

Although polyandry is generally rare among human societies, there are several societies in the highlands of the Himalayas where fraternal polyandry is the preferred form of marriage. In these societies, several brothers marry one woman and establish a communal household. Since polyandry seems to limit a man's reproductive opportunities, we might ask why men tolerate polyandrous marriages. Nancy Levine of the University of California, Los Angeles, studied one of these polyandrous societies, the Nyinba of northwestern Nepal (Figure 19.6). Her data shed some interesting light on this question.

FIGURE 19.6

Polyandry is the preferred form of marriage among the Nyinba. In this family, three brothers are married to one woman. The Nyinba consider this to be the ideal number of co-husbands. (Photograph courtesy of Nancy Levine.)

The Nyinba live in four prosperous villages, nestled between 2850 and 3300 m (9500 and 11,000 ft) on gently sloping hillsides. Nyinba families support themselves through a combination of agriculture, herding, and long-distance trade. The Nyinba believe that marriages of three brothers are most desirable, allowing one husband to farm, another to herd livestock, and the third to engage in trade. In practice, however, all brothers marry jointly, no matter how many there are. The Nyinba are quite concerned with the paternity of their children. Women identify the fathers of each of their children, a task made easier by the fact that one or more of their husbands is often away from home for lengthy periods tending to family business. Although it is impossible to be certain how accurate these paternity assessments are, it is clear that the Nyinba place great stock in them. Men develop particularly close relationships with the children they have fathered, and fathers bequeath their share in the family's landholdings to their own sons.

Evolutionary theory helps to explain why some polyandrous marriages are successful and others fail.

Polyandry is the ideal form of marriage among the Nyinba, but reality does not always conform to this ideal. While all men marry jointly, some marriages break down when one or more of the brothers brings another wife into the household or leaves to set up an independent household. When households break up, each man receives a share of the estate and takes with him all the children he has fathered during the marriage. Levine reasoned that a comparison of marriages that remain intact and marriages that break up would provide clues about the kinds of problems that arise in polyandrous marriages. Levine together with one of us (J. B. S.) has used the data to determine whether male decisions conform to predictions derived from evolutionary theory.

All other things being equal, the more men who are married to a single woman, the lower each man's reproductive success is likely to be. Thus, we might expect marriages with many co-husbands to be less stable than marriages with fewer co-husbands. This turns out to be the case; few of the largest communal Nyinba marriages remained intact. Only 10% of the marriages that began with two co-husbands partitioned, while 58% of the marriages that began with four co-husbands did (Figure 19.7).

FIGURE 19.7
Among the Nyinba, marriages that contain more than three brothers are more likely to break up than marriages that include only two or three co-husbands. Polyandrous households are those that maintain intact polyandrous marriages. Conjoint households are those that added a second wife, which often effectively created two separate marriages within a single household. Partitioned households are those in which one or more men left the fraternal marriage and formed a new household.

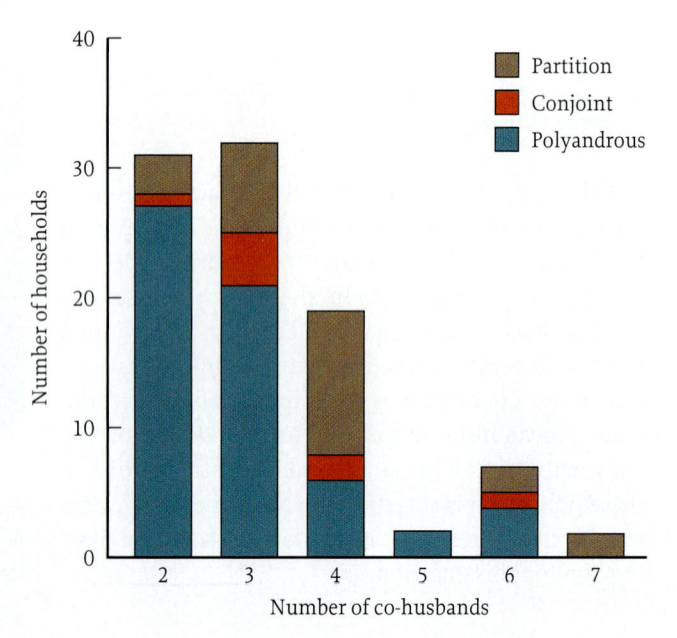

Some Nyinba marriages are arranged by parents, while others are set up by the oldest brother. In either case, the wife is usually a few years younger than the oldest co-husband. In some marriages, the youngest co-husbands are considerably younger than their wives. Remember that Buss's cross-cultural data suggest that men prefer to marry younger women. Thus, we might expect that men would be dissatisfied with marriages to women much older than themselves. In fact, the Nyinba men who were most junior to their wives were the ones most likely to initiate partitions. When these men remarried, they invariably chose women who were younger than themselves and who were younger than their first wives.

If the basic problem with polyandry is that it limits male reproductive opportunities, then we should expect males' satisfaction with their marriages to be linked to their reproductive performance. In most marriages, the oldest brother fathers the wife's first child. The second-oldest brother often fathers the second child, and so on. However, disparities in reproductive success among co-husbands often persist. A man's place in the birth order is directly related to his reproductive success, and the oldest brothers have more children than the youngest brothers. Men's decisions to terminate their marriages are associated with their reproductive performance; men who remained in stable polyandrous marriages fathered 0.1 children per year, while men who terminated their marriages had produced only 0.04 children per year during the relationship.

Finally, kin selection theory would lead us to predict that close kinship among co-husbands would help to stabilize polyandrous marriages because men should be more willing to invest in their brothers' children. We can test this prediction because there is a considerable amount of variation in the degrees of relatedness among co-husbands within households. Some households are composed entirely of full siblings. In others, the co-husbands may have multiple fathers, who are related themselves, and multiple mothers, who may also be related. This produces a complicated web of relationships among co-husbands. Contrary to predictions based on kin selection, there is no difference in the degree of relatedness among co-husbands in households that remained intact and households that dissolved. Moreover, when men remarried, they showed no inclination to align themselves with the co-husbands to whom they were more closely related.

Evolutionary theory explains certain aspects of polyandry but not others.

Polyandry is rare among human societies, as we would expect from an evolutionary perspective. When marriages dissolve, they seem to do so in ways that are consistent with predictions derived from evolutionary theory. Men leave marriages in which the women are much older, and they leave marriages when they have not fathered many children. However, men seem to ignore kinship to their co-husbands in their decisions about whether to maintain their marriages, a surprising result from an evolutionary perspective.

Raising Children

Like other mammalian females, women are predisposed to invest heavily in their offspring. In humans, pregnancy lasts nine months; in traditional societies, lactation lasts at least one year and often longer. Parental investment extends well beyond weaning, as children do not become fully independent until they reach their teens. In most human societies, both mothers and fathers help to raise their own children. All this seems so normal and so obvious that it is easy to overlook the possibility that societies could be organized in many other ways. For example, babies could be bought and sold like pets. Such a system would have lots of practical advantages. It would be easy to adjust the size of your family, to select the ratio or order of boys and girls, and to regulate the spacing between children. Women who preferred not to undergo pregnancy and delivery, or who were unable to bear children, could have a family just as easily as anyone else. Some people might even opt to skip the dubious pleasures of frequent diaper changes and midnight feedings, and acquire children who were two or three years old. But there are no societies in which this happens. Instead, in virtually every society, most people raise their own children. It seems likely that people do this because their evolved psychology causes them to value their own children very differently from other people's children.

There is, however, much more cross-cultural variation in human parenting behavior than in the parenting behavior of other primate species. In some societies, most men are devoted fathers, while in others men take little interest in children and rarely interact with their own offspring. In some societies, parents supervise their children's every waking moment, while in other societies, children are mainly reared by older siblings and cousins. Moreover, some aspects of parenting behavior in some societies seem hard to reconcile with the idea that human psychology has been shaped by natural selection. There are cultures where parents sometimes kill or physically abuse their own infant children, and cultures where people regularly raise children who are not their own. It is also common for people to intentionally limit their own fertility, and to produce fewer children than they are capable of raising. Because such behaviors seem inconsistent with the idea that human behavior is controlled by evolved predispositions, their existence has been cited as evidence that evolutionary reasoning has little relevance to understanding modern human behavior.

In the remainder of this chapter we consider data suggesting that observed patterns of infanticide, child abuse, adoption, and family planning are consistent with the idea that human behavior is influenced by evolved predispositions. Moreover, evolutionary reasoning can give fresh and useful insights about when and why such behaviors occur. However, we will also see that there are aspects of contemporary parenting behavior that evolutionary reasoning has not yet explained.

Child Abuse and Infanticide

Evolutionary theory predicts that parents should terminate investment in offspring if their prospects are poor.

Children have been deliberately killed, fatally neglected, or abandoned by their parents through recorded history and across the world's cultures. It is tempting to dismiss infanticide by parents as a pathological side effect of modern life. After all, a child is a parent's main vehicle for perpetuating his or her own genetic material. However, evolutionary analyses suggest that parents who kill their own children may, under some circumstances, have higher fitness than parents who do not do so. Natural selection may have molded our psychology so that we are sometimes prone to harm or neglect children under very special circumstances. If this assertion seems preposterous, remember from Chapter 17 that natural selection favors physiological mechanisms that terminate pregnancy when fetal prospects for survival are poor. The same adaptive logic applies to infants after they are born. Evolutionary reasoning suggests that parents should be predisposed to care only for offspring that are likely to survive and to reproduce successfully. Although this may seem cold, calculating, and cruel, it must be viewed in the context of human evolutionary history. In every plausible environment of evolutionary adaptedness, parental ability to invest in their offspring would have been limited, and natural selection would have favored psychological mechanisms that cause parents to terminate investment in offspring that are not likely to survive, and to channel investment toward the children that are most likely to live to maturity and to reproduce successfully. We expect these psychological mechanisms to be sensitive to the child's condition, the parents' economic circumstances, and the parents' alternative opportunities to pass on their genetic material. This phenomenon has been labeled **discriminative parental solicitude** by psychologists Martin Daly and Margo Wilson of McMaster University in Ontario, Canada.

Cross-Cultural Patterns of Infanticide

Cross-cultural analyses indicate that infanticide occurs when a child is unlikely to survive, parents cannot care for the child, or the child is not sired by the mother's husband.

Daly and Wilson used the Human Relations Area Files, a vast compendium of ethnographic information from societies around the world, to compile a list of the reasons why parents commit infanticide. They based their study on a standard, randomly selected sample of 60 societies that represent traditional societies around the world. In 39 of these societies, infanticide is mentioned in ethnographic accounts, and in 35 of the societies Daly and Wilson found information about the circumstances under which infanticide occurs. There are three main classes of reasons why parents commit infanticide: 1) the child is seriously ill or deformed, 2) the parents' circumstances do not allow them to raise the child, or 3) the child is not sired by the mother's husband (Table 19.2).

In 21 societies, children are killed if they are born with major deformities or if they are very ill. In traditional societies, children with major deformities or serious illnesses require substantial amounts of care and impose a considerable burden on their families. They are unlikely to survive, and if they do, they are unlikely to be able to support themselves, marry, or produce children of their own. Thus, it is plausible that selection has shaped human psychology so that we are predisposed to terminate investment in such children and to reserve resources for healthy children.

In some societies, children are killed when the parents' present circumstances make it too difficult or dangerous to raise them. This includes cases in which births are too closely spaced, twins are born, the mother has died, there is no male present to support the children, or the mother is unmarried. It is easy to see why these factors might make it very difficult to raise children. Twins, for example, strain a mother's abilities to nurse her children and to provide adequate care for them. Moreover, twins often are born prematurely and are of below-average birth weight, making them less likely to survive than single infants. In traditional societies, it is not always possible to obtain substitutes for breast milk. This dooms some children whose mothers die when they are very young. If a child's father dies before it is born or when it is an infant, its mother may have great difficulty supporting herself and her older children. The child's presence might also reduce her chances of remarrying and producing additional children in the future. When circumstances are not favorable, parents who attempt to rear children may squander precious resources that might better be reserved for older children or for children born at more favorable times.

TABLE 19.2 In a representative cross-cultural sample of societies, the rationales for infanticide are consistent with predictions derived from evolutionary theory. Parents mainly commit infanticide when the child is unlikely to thrive, when their own circumstances make it difficult to raise the child, or when the child was not fathered by the mother's spouse. (Data from Table 3.1 in M. Daly and M. Wilson, 1988, *Homicide,* Aldine de Gruyter, New York.)

Rationale	Number of Societies
Poor quality of child	
Deformed or ill	21
Parents' circumstances do not allow them to raise child	
Twins born	14
Children too closely spaced or numerous	11
No male support for child	6
Mother died	6
Mother unmarried	14
Economic hardship	3
Born in wrong season	1
Quarrel with husband	1
Paternity not assigned to wife's husband	
Adulterous conception	15
Nontribal sire	3
Sired by mother's previous husband	2
Other reasons	15

In 20 societies, infanticide occurs when a child is sired during an extramarital relationship, by a previous husband, or by a member of a different tribe who is not married to the mother. Recall from Chapter 8 that the theory of kin selection predicts that altruism will be restricted to kin. Males should not be predisposed to support other men's children.

Before you condemn those who kill their children, keep in mind that the options for parents in traditional societies are very different from the options available to people in our own society who have access to modern health-care systems, social services, and so on. Decisions by parents to kill their children are often reluctant responses to the painful realities of life. Whether these considerations are sufficient to absolve the parents in your eyes depends, of course, on your moral beliefs. However, there is little doubt that parents usually deeply regret the need to kill a newborn infant and may grieve over the loss for many years.

Child Abuse in the United States and Canada

Evolutionary reasoning predicts that child abuse in North America will occur under the same circumstances that infanticide occurs cross-culturally.

Richard Gelles of the University of Rhode Island, a national expert on child abuse, estimates that at least 1 million children are physically, emotionally, or sexually abused in the United States each year. Since many instances of abuse are not reported to authorities, the actual incidence of abuse may be much higher. It seems unlikely that child abuse in our own society is actually adaptive in the sense that it increases the abusive parent's reproductive success. Very few North Americans face the harsh necessity of killing a child in order to allow others to survive. However, it is plausible that the same psychological mechanisms that lead to infanticide in simpler, more resource-constrained societies may lead to child abuse in our own society. Evolutionary theory predicts that people have an evolved predisposition to cherish, protect, and provide for their own children. However, it also predicts that these bonds will weaken when families are stressed, when the child is sickly or handicapped, or when the child's paternity is uncertain. Evolutionary reasoning predicts that child abuse is most likely to occur in these circumstances.

Martin Daly and Margo Wilson applied this reasoning to an analysis of the factors that lead to child abuse in industrialized societies like the United States. Detailed, quantitative information about child abuse was difficult to come by when they began their work. They pored through descriptions of child homicides in police files, and combed statistical reports on child abuse maintained by governmental agencies and private organizations. They looked for records that would provide information relevant to evolutionary hypotheses about the conditions that lead to mistreatment and abuse of children, such as current parental circumstances and the abused child's relationship to members of its household. In the end, they were able to draw several important conclusions from the data.

Children living in households with a biological parent and an unrelated adult are at much higher risk of child abuse than are children living with both biological parents.

Cross-cultural research indicates that infanticide often occurs when a child is sired by someone other than the mother's current husband. We have seen that this makes sense from an evolutionary point of view, since men are not expected to invest in other men's children. Similar factors influence the risk of child abuse in Western industrialized societies. Due to divorce and other factors, a substantial number of children in the United States and other industrialized nations live in households with an unrelated **substitute parent,** such as a stepparent. Adoptive parents are not included in this category. Daly and Wilson, and others, have found that the risk of abuse is much lower in households that include two biological parents than in households that contain one natural parent and a stepparent. The data in Figure 19.8 come from Canada in 1983, but data from the United States, a much more violent country than Canada, show essentially the same pattern.

Evolutionary reasoning predicts that the substitute parent is usually responsible for the child abuse.

The statistics are silent about the identities and motives of the individuals responsible for abuse. However, evolutionary reasoning implicates the substitute parents. If, as Daly and Wilson suggest, evolution has predisposed us to form bonds with the children that we know intimately as infants, then it may be difficult for substitute parents to establish loving relationships with their partner's children. Unlike most biological parents and virtually all adoptive parents, substitute parents may be unprepared or reluctant to take on the responsibilities and expenses associated with parenthood. They may find their partner's children noisy, demanding, and irritating, and resent the time and money that their partner devotes to them. Even the best-intentioned substitute parents may find it difficult to establish loving bonds with their partner's children, particularly if the substitute comes on the scene late in the child's life.

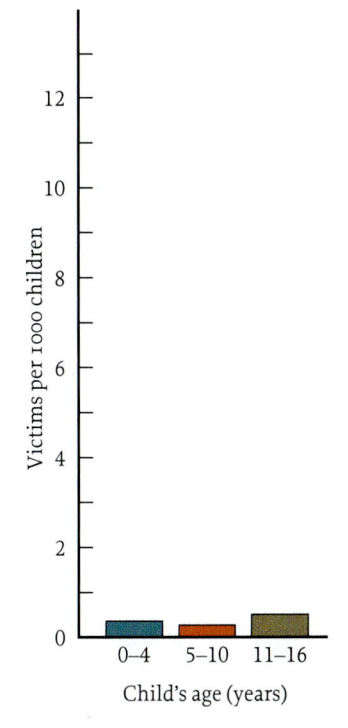

(a) Two natural parents

(b) One natural and one stepparent

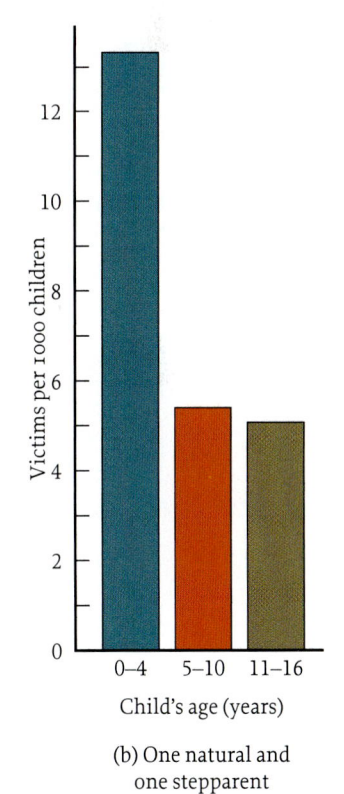

FIGURE 19.8

(a) In Canada, children living in households with two natural parents are at lower risk for child abuse than (b) children living with one natural parent and one stepparent, at all ages.

Adoption

In many societies children are adopted and raised by adults other than their biological parents.

Adoption is the flip side of infanticide. In some regions of the world, particularly Oceania and the Arctic, a substantial fraction of all children are raised in adoptive households. Adoption can be thought of as a form of altruism, because it takes considerable time, energy, and resources to raise children. In other animal species, including other primates, voluntary extended care of others' offspring is uncommon. There are, however, examples of *in*voluntary extended care: some species of parasitic birds, like cuckoos, routinely lay their eggs in other birds' nests, tricking the unwitting hosts into rearing cuckoo young as their own (Figure 19.9). The fact that we label such behavior **nest parasitism** reflects the fact that this is a form of exploitation, enhancing the fitness of the parasitic cuckoo at the expense of its host. In human societies, there is no need to trick adoptive parents into caring for children, as they are typically eager to assume responsibility for other people's children.

FIGURE 19.9
Cuckoos lay their eggs in the nests of other species of birds. The hosts unwittingly rear the alien chicks as their own.

Adoption in Oceania

Adoption is very common in the societies of the Pacific Islands. The pattern of adoption in those societies is consistent with the predictions of evolutionary theory.

At first glance, adoption seems inconsistent with evolutionary theory, which predicts that people should be reluctant to invest time and energy in unrelated children. This apparent inconsistency prompted Marshall Sahlins, a cultural anthropologist at the University of Chicago, to cite adoption as an example of a human behavior that contradicts the logic of evolutionary theory. In *The Use and Abuse of Biology*, Sahlins pointed out that adoption is very common in the societies of the Pacific Islands. In fact, it is so common that the majority of households in many of these island societies include at least one adopted child. According to Sahlins, such widespread altruism toward nonkin shows that evolutionary reasoning does not apply to contemporary humans, and demonstrates that human societies are free to invent almost any social arrangements.

Sahlins's challenge prompted one of us (J. B. S.) to take a closer look at the pattern of adoption in a number of the societies of Oceania. The data tell a very interesting story. Adoption seems to provide an adaptive means for birth parents to regulate the sizes of their families and to enhance the quality of care they give to each of their offspring. Moreover, many of the features of adoption transactions in Oceania are consistent with predictions derived from kin selection theory:

- Adoptive parents are usually close kin, such as grandparents, aunts, or uncles (Figure 19.10). Close kin participate in adoptive transactions much more often than would be expected if children were adopted at random, without respect to kinship.
- Natural parents give up children when they cannot afford to raise them, and they rarely give up firstborn children for adoption.

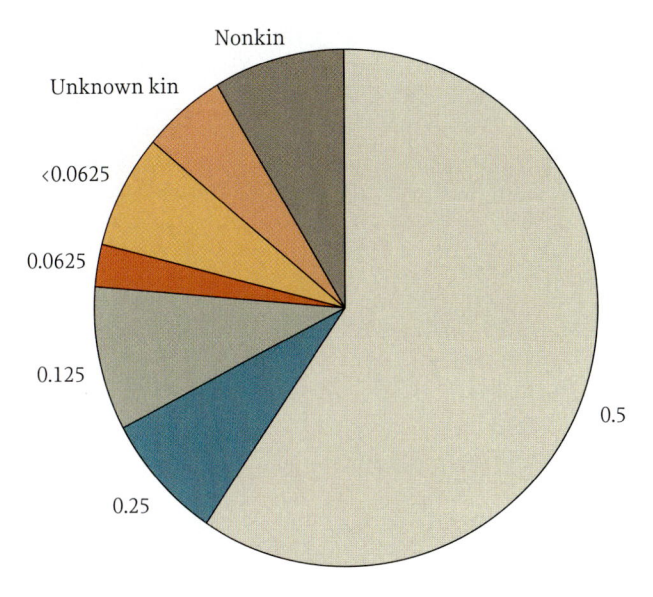

FIGURE 19.10

People in Oceania mainly adopt kin. Information about the numbers of adoptive parents and their relationships with their adoptive children in a number of communities in Oceania have been collected by ethnographers. The numbers represent the degree of relatedness between the child and its adoptive parents. Most adoptions involve close kin ($r \geq .25$), such as aunts, uncles, and grandparents.

- Adoptive parents generally have no dependent children; they are usually childless or the parents of grown children, but sometimes they are simply wealthy enough to be able to raise an additional child.
- Natural parents are often reluctant to give up their children for adoption, and regret the need to do so.
- Natural parents maintain contact with their children after adoption, and terminate adoptions if children are neglected, mistreated, or unhappy.
- Natural parents prefer to have their children adopted by well-to-do individuals who can provide adequately for their daily needs or bestow property on them.
- Sometimes asymmetries in investment exist between natural and adoptive children, as adopted children often inherit less property from their adoptive parents than the biological offspring do.

The same pattern characterizes adoptions in some traditional societies elsewhere.

Many of the same features characterize adoption in the North American Arctic, where many children are adopted. Moreover, adoption transactions in Oceania and the North American Arctic bear a striking similarity to more temporary fosterage arrangements common in West Africa. In each of these cases, birth parents seem to act in ways that increase the health, security, and welfare of their children, and care for children is preferentially delegated to close kin.

Adoption transactions seem to fit predictions derived from the theory of kin selection, but this does not mean that we have fully explained human adoption transactions. Not all adoption transactions in Oceania fit the evolutionary model. We have not accounted for the fact that adoption is more common in some societies than in others, or that there are some societies, including our own, in which adoption is not restricted to relatives. It is to this problem we now turn.

ADOPTION IN INDUSTRIALIZED SOCIETIES

In contemporary industrial societies, adoption often involves strangers.

As you read the description of adoption in the traditional societies of Oceania, you may have been struck by how different these transactions are from adoptions in the

TABLE 19.3 The patterns of adoption in the United States and Oceania show both similarities and differences.

Feature	Oceania	United States
Economic difficulties figure in the adoption decision.	Yes	Yes
Adoptive parents are usually childless or wealthy enough to raise another child.	Yes	Yes
Children are adopted by close kin.	Usually	Sometimes*
Biological parents often regret the need for adoption.	Yes	Yes
Legal authority for the child is transferred exclusively to the adoptive parents.	No	Yes
The identities of the natural parents are known to the adoptee.	Yes	No
Contact is maintained between biological parents and child after adoption.	Yes	No
Biological parents want children to be adopted by well-to-do people.	Yes	Yes
Asymmetries exist in the care of adoptive children and biological children.	Yes	?

*In the United States from 1952 to 1971, approximately one-fourth of adoptions involved kin, another fourth involved stepparents, and one-half involved unrelated individuals.

United States. In the United States and in other industrialized nations, adoption is often a formal, legal process involving strangers. However, as you compare the features of adoption transactions in Oceania and the United States in Table 19.3, you will see that there are similarities as well as differences between them.

In Oceania and in the United States, children are given up for adoption mainly when their parents are unable to raise them, and they are frequently adopted by people who have no dependent children of their own or by people who can afford to raise additional children. In both Oceania and the United States, biological parents often give up their children with considerable reluctance and hope to place their children in homes where their prospects will be improved.

The main difference between adoption in Oceania and in the United States is that in Oceania, adoption is a relatively informal, open transaction between close kin, while in the United States adoption is often a formal, legal transaction between strangers. This has a number of ramifications. When the identity of the biological parents is not disclosed to the adoptive parents or the adopted child, contact between biological parents and their children is broken. In Oceania, adopted children's interests are protected by their biological parents, who maintain contact with them and reserve the right to terminate the adoption if the children are mistreated or unhappy. In the United States, governmental agencies are responsible for protecting the interests of adopted children. In recent years, some of these differences have been eroded as open adoptions have become more common, and the courts have given adoptees the right to find out the identities of their birth parents.

The different patterns of adoption in Oceania and the United States may result from the same basic evolved psychological motivations.

Overall, it seems that the similarities in adoption are related to people's motivations about children; the differences are related to how the transactions are organized in each culture. In Oceania and the United States, people seem to share concerns about their children. We are deeply concerned about the welfare of our own children, and many people have deep desires to raise children. These feelings are likely to be the product of evolved psychological predispositions that motivate us to cherish and protect children. Although people may have very similar feelings about children in Oceania and the United States, the adoption process is clearly different. Nepotism (favoring relatives over nonrelatives) is a central element in adoption transactions in Oceania but not in the United States. People in Oceania rely on their families to help them find homes for their children and to obtain children, while people in the United States turn to adoption agencies, private attorneys, and strangers.

We are not sure why kinship plays such a fundamental role in adoption transactions in Oceania, but not in industrialized societies like the United States and Canada. We can speculate that it is due to differences in the availability of children who are eligible for adoption, or to the tendency of family members to be dispersed geographically. However, nepotism generally seems to play a more important role in the societies of Oceania than it does in industrialized societies.

Family Size

!Kung women deliberately space their children's births at long intervals and thus seem to have fewer children than they could during their reproductive years.

As we have repeatedly emphasized, natural selection favors adaptations that increase an individual's reproductive success. Parents are expected to produce as many offspring as they can. In many foraging societies, however, children are spaced carefully and born at relatively long intervals. The Dobe !Kung, one of the best-studied foraging groups, provide a good example. (Dobe is a place on the Botswana-Namibia border, and !Kung is a language group.) !Kung women give birth to their first child in their late teens and their last child in their mid-40s. The interval between births of surviving children is usually four years (Figure 19.11). Thus, during their 20-year reproductive careers, women give birth to about five children. If interbirth intervals were shortened, women would be able to have more children. From an evolutionary perspective, we must ask why the Dobe !Kung have children at such long intervals.

!Kung women space their births to avoid having to carry more than one young child.

The lives of Dobe !Kung women are described in rich detail by Richard Lee of the University of Toronto. The Dobe !Kung live in the vast reaches of the Kalahari Desert and subsist on a variety of plant foods and wild game. In this inhospitable terrain (Figure 19.12), women work hard to provide food for themselves and their families. The bulk of the calories in the Dobe !Kung diet come from plant foods gathered mainly by the women. During the driest part of the year, the Dobe !Kung camp near a few widely spaced water holes. The women travel long distances in search of food, foraging in mongongo nut groves that may be up to 10 km (6 miles)

FIGURE 19.11

Among the !Kung, children are spaced about four years apart. This woman is pregnant and is flanked by her husband and two children.

FIGURE 19.12
The !Kung now live in the Kalahari Desert, a harsh and inhospitable habitat. (Photograph courtesy of Nicholas Blurton Jones.)

FIGURE 19.13
Women carry heavy loads of nuts and other plant foods back to camp to share with their families. (a) A woman fills her sack with nuts, and then (b) her toddler climbs on top. (Photographs courtesy of Nicholas Blurton Jones.)

from their camps and water holes. The Dobe !Kung do not have donkeys or horses, so women forage on foot and carry loads on their backs. When women go out to forage, they take their children with them. They carry their babies for the entire journey, and children up to the age of four need to be carried at least part of the way (Figure 19.13). Children obtain very little of their own food until they reach their teens. Some of the children's calories come from plant foods gathered by their mothers, and some from food brought back and shared by hunters. Overall, women provide about 60% of their children's calories until the offspring reach the age of 15 years.

Nicholas Blurton Jones of the University of California, Los Angeles, has tried to determine why the !Kung have so few children. Following up on Lee's observation that carrying children is very hard work, Blurton Jones hypothesized that for Dobe !Kung women, interbirth intervals may be limited by the weight of the loads they can carry on their backs. Since children cannot walk long distances until they are four years or older, women who had children at short intervals would have to carry more than one child. One way for the women to restrict their loads to manageable limits would be to extend their interbirth intervals.

The dynamics of the problem are portrayed in Figure 19.14. The graph shows that the load a woman must carry has two components: the weight of the child and the weight of the food the child eats. From the time of a child's birth until the child is four years old, the weight of the mother's load increases, because the child gets heavier as it gets older. As the child matures, it begins walking longer distances on its own, thereby lightening the mother's load. Of course, as children grow, they need more food. Hence, the weight of food carried back for children increases as they get older. However, children weigh considerably more than the food required to sustain them each day. By combining the weight of the child with the weight of the food carried back to feed the child, Blurton Jones calculated the total weight of the load that a woman must carry to support her child. A woman with a two-year-old child carries about 7 kg (15 lb), while a woman with a mobile six-year-old child carries less than 2.25 kg (about 5 lb).

These data can then be used to compare the weight of women's loads if they have children at different intervals (Figure 19.15). When women give birth at short intervals (such as every two years), their loads are very heavy (more than 22.5 kg, or 50 lb) because they must carry two children, along with the food needed to feed them both. Women who give birth at longer intervals have lighter loads because their older children don't need to be carried as much of the time. It is important to notice that while women's loads become steadily lighter as interbirth intervals are extended to four years, the loads do not become appreciably lighter after this point. Thus, women may space their children at four-year intervals because this is the spacing that minimizes women's loads and maximizes their reproductive rate. The average interbirth interval is four years, just as the model predicts. Dobe !Kung women are actually having children as often as they can, given the immense effort associated with child care.

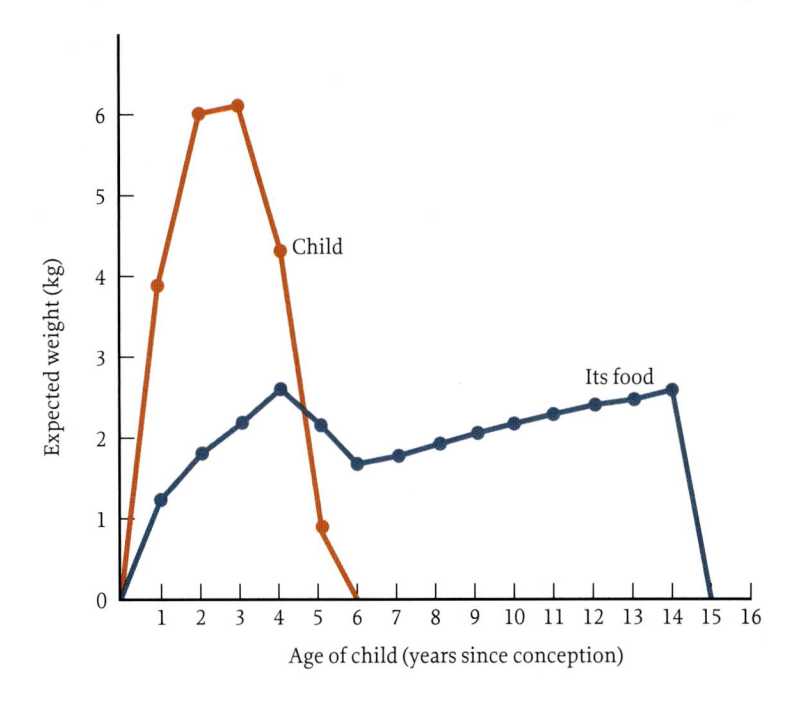

FIGURE 19.14

Loads carried by !Kung women decrease with the child's age. There are two components to the mother's load: the weight of the child (plotted in red) and the weight of the child's food (plotted in blue).

Blurton Jones's model is supported by the fact that !Kung women who do not need to forage have shorter interbirth intervals.

One prediction of the "back-load" model is that when a Dobe !Kung woman's workload is reduced, her interbirth intervals will decline. There is evidence that this is the case. The data represented in Figures 19.14 and 19.15 are based on women who live in the bush and rely on foraging for subsistence. However, some Dobe !Kung families

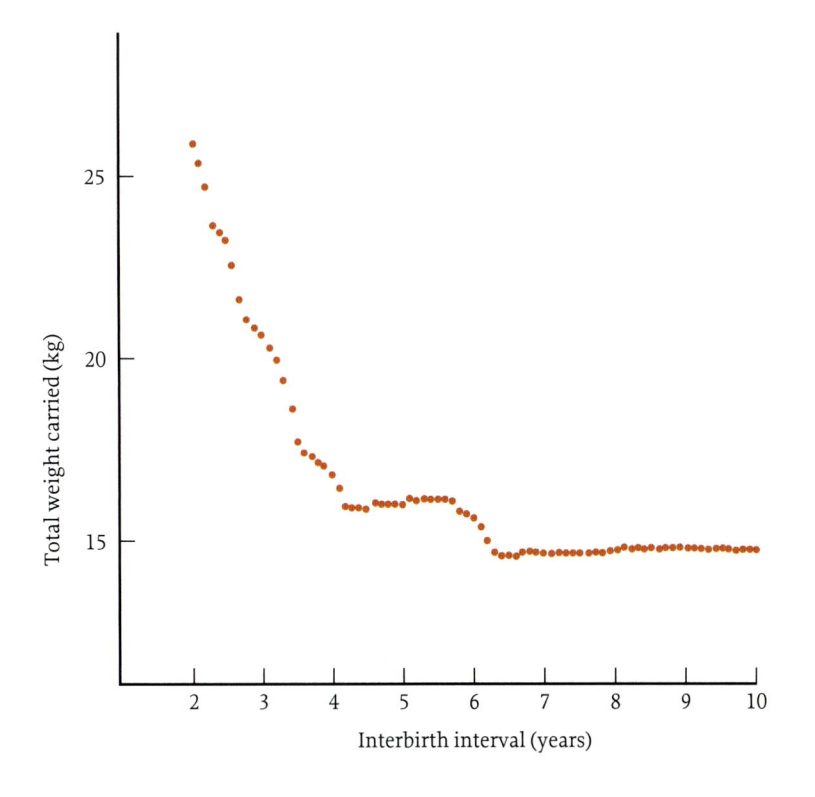

FIGURE 19.15

The total load carried by !Kung women is minimized when births are spaced at approximately four-year intervals. When interbirth intervals are shorter, women are forced to carry heavy loads. When interbirth intervals are longer than four years, women do not appreciably lighten their load.

have settled at cattle posts (as cattle-raising operations are called in southern Africa) where they work for their Herero and Tswana neighbors. Women who live at the cattle posts do not forage in the bush as often and do not have to carry their children over long distances. As the model would predict, Dobe !Kung women living at cattle posts have much shorter interbirth intervals than foraging women. Another implication of the back-load model is that women who live in the bush and have children at shorter intervals will reproduce less successfully than women who have children at longer intervals. In the bush, infants born after relatively short interbirth intervals are less likely to survive than those born after longer intervals.

This evolutionary explanation of birth spacing leaves unexplained many factors that affect birth spacing.

The analysis of !Kung birth spacing takes as given many features of !Kung society described by Lee and others. For example, it assumes that children do not acquire very much of their own food, that men don't contribute very much food to the family pot, and that women don't cooperate more in caring for each other's children. If any one of these features were changed, !Kung women might be able to have more children. Moreover, it is known that some of these features vary cross-culturally. For example, Blurton Jones has studied a second group of foragers, the Hadza of Tanzania. Hadza children acquire a much larger amount of their own food than do !Kung children (Figure 19.16). It may be that each of these features has an adaptive explanation. For example, Blurton Jones has suggested that Hadza children forage more than !Kung children do because there is much more food in the immediate vicinity of Hadza camps. On the other hand, these features may be maintained as the result of culturally transmitted norms, and people with a different cultural history might behave quite differently under the same conditions.

FIGURE 19.16
Hadza children gather much of their own food from an early age, perhaps because they can find food close to their camps. Here several children draw water out of cracks in the rocks. (Photograph courtesy of Nicholas Blurton Jones.)

Is Human Evolution Over?

This question is often raised by students in our courses, and it seems to be a sensible question to consider as we come to the end of the story of human evolution. As we have seen, modern humans represent the product of millions of years of evolutionary change. Of course, so do cockroaches, peacocks, and orchids. All of the organisms that we see around us, including people, are the products of evolution, but they are not finished products. They are simply works-in-progress.

There is a sense, however, in which human evolution is over. Because cultural change is much faster than genetic change, most of the changes in human societies since at least the origin of agriculture, almost 10 kya, have been the result of cultural, not genetic evolution. Most of the evolution of human behavior and human societies is not driven by natural selection and the other processes of organic evolution. It is

The Relationship between Science and Morality

Recently, after giving a broadcast on Charles Darwin, I received through the post a pamphlet, 'Why are there "Gays" at all? Why hasn't evolution eliminated "Gayness" millions of years ago?', by Don Smith. The pamphlet points to a genuine problem; the prevalence of sexually ambiguous behaviour in our species is not understood, and is certainly not something that would be predicted from Darwinian theory. Smith's motive for writing the pamphlet was as follows. He believes that the persecution of gays has been strengthened and justified by the existence of a theory of evolution which asserts that gays are of low fitness because they do not reproduce their kind. He also believes that gays can only be protected from future persecution if it can be shown that they have played an essential and creative role in evolution.

I do not find the evolutionary theory he offers in the place of Darwinism particularly persuasive, although it is neither dull nor silly. However, that is not the point I want to make here. I think he would have been better advised to say:

> If people have despised gays because gayness does not contribute to biological fitness, they have been wrong to do so. It would be as sensible to persecute mathematicians because an ability to solve differential equations does not contribute to fitness. A scientific theory—Darwinism or any other—has nothing to say about the value of a human being.

The point I am making is that Smith is demanding of evolutionary biology that it be a myth; that is, that it be a story with a moral message. He is not alone in this. Elaine Morgan recently wrote an account of the origin of *H.sapiens* intended to dignify the role of women and of the mother-child bond (a relationship about which Don Smith is silent). Earlier, Shaw wrote *Back to Methuselah* avowedly as an evolutionary myth, because he found in Darwinism a justification of selfishness and brutality, and in Lamarckism a theory which justified free will and individual endeavour.

We should not be surprised at Don Smith, Elaine Morgan and Bernard Shaw. In all societies men have constructed myths about the origins of the universe and of man. The function of these myths is to define man's place in nature, and thus to give him a sense of purpose and value. Darwinism is, among other things, an account of man's origins. Is it to be wondered at that it is expected to carry a moral message?...

... We do not find it easy to distinguish science and myth. One reaction to this difficulty is to assert that there *is* no difference. Evolution theory has no more claim to objective truth than Genesis. Many scientists would be enraged by such an assertion, but rage is no substitute for argument. In the last century, it was widely held that the scientific method, conceived of as establishing theories by induction from observation, led to certain knowledge. If that were so, then there would indeed be a way to distinguish science from myth, because the truth of a myth certainly cannot be established by induction. However, Darwin and Einstein have robbed us of that confidence in induction—or have liberated us from that prison. By establishing the mutability of species, Darwin showed that there is not a fixed and finite number of kinds of things in the universe, each with a knowable essence; there is no 'Platonic idea' for each species. But induction can only lead to certainty if there is a finite and knowable set of objects, so that one can check that one's theory is true of every kind. If Darwin demonstrated the impossibility of acquiring certainty through induction, Einstein showed that what scientists had been most certain of—classical mechanics—was at worst false and at best a special case of a more general theory. After this twin blow, sure and certain knowledge is something we can expect only at our funerals.

But it is one thing to say that scientific knowledge cannot be certain, and quite another that there is no difference between science and myth....

... Indeed, they have much in common. Both are constructs of the human mind, and both are intended to have a significance wider than the direct assertions they contain.... It is the function of a scientific theory to account for experience—often, it is true, the rather esoteric experience emerging from deliberate experiment. It is the function of a myth to provide a source and justification for values. What should be the relation between them?

Three views are tenable. The first, sometimes expressed as a demand for 'normative science', is that the same mental constructs should serve both as myths and as scientific theories. It is widely held. If I am right, it underlies the criticisms of Darwinism from gays, from the women's movement, from socialists, and so on. It explains the preference expressed by some churchmen for 'big bang' as opposed to 'steady state' theories in cosmology. Although well-intentioned, it seems to me pernicious in its effects. If we insist that scientific theories convey moral messages, the result will be bad morality or bad science, and most probably both. The danger is most apparent in evolutionary biology. Darwinism *is* an account of human origins. In all previous cultures, accounts of origins have been myths, so that it is to be expected that people will treat Darwinism as a myth. If one accepts the Darwinian account, then it is easy to equate 'natural' with 'successful in the struggle for existence'; if one treats the account as a myth, it is equally easy to equate 'natural' with 'right'. The consequence is that people either embrace Darwinism and draw from it the conclusion that gays are unnatural, social services impolitic, and charity wicked, or they are so disgusted by these conclusions that they embrace Lamarckism whether or not the evidence supports it. The first choice is bad morality and the second is bad science. There is no escape from this dilemma, so long as we insist on treating scientific theories as if they were myths. However difficult it may be to convince ourselves and others that 'natural' is *not* equivalent to 'right', the attempts must be made.

The second view is that we should do without myths and confine ourselves to scientific theories. This is the view I held at the age of twenty, but it really won't do. If, as I believe, scientific theories say nothing about what is right, but only about what is possible, we need some other source of values, and that source has to be myth in the broadest sense of the term.

The third view, and I think the only sensible one, is that we need both myths and scientific theories, but that we must be as clear as we can which is which.... To do science, one must first be committed to some values—not least to the value of seeking the truth. Since this value cannot be derived from science, it must be seen as a prior moral commitment, needed before science is possible.

SOURCE: From pp. 39–40, 42–43, and 49–50 in J. M. Smith, 1989, *Did Darwin Get It Right? Essays on Games, Sex and Evolution,* Chapman & Hall, New York.

driven by learning and other psychological mechanisms that shape cultural evolution. However, this fact does not mean that evolutionary theory or human evolutionary history is irrelevant to understanding contemporary human behavior. Natural selection has shaped the physiological mechanisms and psychological machinery that govern learning and other mechanisms of cultural change, and an understanding of human evolution can provide important insights into human nature and the behavior of modern peoples.

Further Reading

Blurton Jones, N. 1989. The costs of children and the adaptive scheduling of births: towards a sociobiological perspective on demography. Pp. 265–282 in *The Sociobiology of Sexual and Reproductive Strategies,* ed. by A. E. Rasa, C. Vogel, and E. Voland. Chapman & Hall, London.

Borgerhoff Mulder, M. 1988. Kipsigis bridewealth payments. Pp. 65–82 in *Human Reproductive Behavior,* ed. by L. Betzig, M. Borgerhoff Mulder, and P. Turke. Cambridge University Press, Cambridge.

Buss, D. 1994. *The Evolution of Desire.* Basic Books, New York.

Daly, M., and M. Wilson. 1988. *Homicide.* Aldine de Gruyter, Hawthorne, N.Y.
Levine, N. E. 1988. *The Dynamics of Polyandry.* University of Chicago Press, Chicago.
Silk, J. B. 1990. Human adoption in evolutionary perspective. *Human Nature* 1:25–52.

Study Questions

1. In Chapter 7 we said that the reproductive success of most male primates depends on the number of females with which they mate. Here Buss argues that a man's reproductive success will depend mainly on the health and fertility of his mate. Why are humans different from most primates? Among what other primate species should we expect males to attend to the physical characteristics of females when choosing mates?
2. Why should men value fidelity in prospective mates more than women do?
3. In Buss's cross-cultural survey, what was the most important attribute in a mate for both men and women? Does this result falsify his evolutionary reasoning?
4. Are Borgerhoff Mulder's observations about the Kipsigis consistent with Buss's cross-cultural results? Explain why or why not.
5. Explain why polyandry is so rare among mammalian species.
6. Explain why individual organisms may sometimes increase their fitness by killing their own offspring. What does your reasoning predict about the contexts in which infanticide will occur?
7. Newborn babies attract an incredible amount of attention. Complete strangers stop to chuck them under the chin, coo at them, and comment on how cute they are. If natural selection has favored discriminative parental solicitude, as the data on infanticide and child abuse seem to suggest, how can you explain the unselective attraction to newborn infants?
8. Why is adoption altruistic from an evolutionary perspective? How can adopting other people's babies be adaptive?
9. Explain why !Kung women have so few children.

EPILOGUE

There Is Grandeur in This View of Life . . .

Here we end our account of how humans evolved. As we promised in the Prologue, the story has not been a simple one. We began, in Part One, by explaining how evolution works—how evolutionary processes create the exquisite complexity of organic design, and how these processes give rise to the stunning diversity of life. Next, we used these ideas in Part Two to understand the ecology and behavior of nonhuman primates—why they live in groups, why the behavior of males and females differs, why animals compete and cooperate, and why primates are so smart compared with other kinds of animals. Then in Part Three we combined our understanding of how evolution works and our knowledge of the behavior of other primates with information gleaned from the fossil record to reconstruct the history of the human lineage. We traced each step in the transformation from a shrewlike insectivore living at the time of dinosaurs; to a monkeylike creature inhabiting the Oligocene swamps of northern Africa; to an apelike creature living in the canopy of the Miocene forests; to the small-brained, bipedal hominids who ranged over Pliocene woodlands and savannas; to the large-brained and technically more skilled early members of the genus *Homo,* who migrated to most of the Old World; and finally, to creatures much like ourselves who created spectacular art, constructed simple structures, and hunted large and dangerous game just 100 kya. In the final part of the book, we turned to look at ourselves—to assess the magnitude and significance of genetic variation in the human species, to ask why we grow old and die, and to consider how we choose our mates and how we raise our children.

Evolutionary analyses of human behavior are not always well received. In Darwin's day, many were deeply troubled by the implications of his theory. One Victorian matron, informed that Darwin believed humans to be descended from apes, is reported to have said, "Let us hope that it is not true, and if it is true, that it does not become widely known." Darwin's theory profoundly changed the way we see ourselves. Before Darwin, most people believed that humans were fundamentally different from other animals. Human uniqueness and human superiority were unquestioned. But we now know that all aspects of the human phenotype are products of organic evolution, exactly the same processes that create the diversity of life around us. Nonetheless, many people still feel that we diminish ourselves by explaining human behavior in the same terms we use to explain the behavior of chimpanzees or soapberry bugs or finches.

FIGURE 1
Charles Darwin died in 1882 and was buried in Westminster Abbey beneath the monument to Isaac Newton.

In contrast, we think the story of human evolution is breathtaking in its grandeur. With a few simple processes, we can explain how we arose, why we are the way we are, and how we relate to the rest of the universe. It is an amazing story. But perhaps Darwin himself put it best in the final passage of *On the Origin of Species:*

> It is interesting to contemplate an entangled bank, clothed with many plants of many kinds, with birds singing on the bushes, with various insects flitting about, and with worms crawling through the damp earth, and to reflect that these elaborately constructed forms, so different from each other, and dependent on each other in so complex a manner, have all been produced by laws acting around us. These laws, taken in the largest sense, being Growth with Reproduction; Inheritance which is almost implied by reproduction; Variability from the indirect and direct action of the external conditions of life, and from use and disuse; a Ratio of Increase so high as to lead to a Struggle for Life, and as a consequence, Natural Selection, entailing Divergence of Character and the Extinction of less-improved forms. Thus, from the war of nature, from famine and death, the most exalted object which we are capable of conceiving, namely, the production of the higher animals, directly follows. There is grandeur in this view of life, with it several powers having been originally breathed into a few forms or only one; and that, whilst this planet has gone cycling on according to the fixed law of gravity, from so simple a beginning endless forms most beautiful and most wonderful have been, and are being evolved [From p. 490 in C. Darwin, 1964 (1859), *On the Origin of Species,* 1st ed., Harvard University Press, Cambridge, Mass.].